GOD, CHANCE AND PURPOSE

Scientific accounts of existence give chance a central role. At the
smallest level, quantum theory involves uncertainty, and evolution
is driven by chance and necessity. These ideas do not fit easily with
theology, in which chance has been seen as the enemy of purpose.
One option is to argue, as proponents of Intelligent Design do, that
chance is not real and can be replaced by the work of a Designer.
Others adhere to a deterministic theology in which God is in total
control. Neither of these views, it is argued, does justice to the
complexity of nature or the greatness of God. The thesis of this book
is that chance is neither unreal nor non-existent but an integral part
of God's creation. This view is expounded, illustrated and defended
by drawing on the resources of probability theory and numerous
examples from the natural and social worlds.

DAVID J. BARTHOLOMEW is Emeritus Professor of Statistics at
the London School of Economics and Political Science. His numerous publications include *Measuring Intelligence: Facts and Fallacies*
(2004).

GOD, CHANCE AND PURPOSE

Can God Have It Both Ways?

DAVID J. BARTHOLOMEW

CAMBRIDGE
UNIVERSITY PRESS

CAMBRIDGE UNIVERSITY PRESS
Cambridge, New York, Melbourne, Madrid, Cape Town, Singapore, São Paulo, Delhi

Cambridge University Press
The Edinburgh Building, Cambridge CB2 8RU, UK

Published in the United States of America by Cambridge University Press, New York

www.cambridge.org
Information on this title: www.cambridge.org/9780521707084

First published 2008

Printed in the United Kingdom at the University Press, Cambridge

A catalogue record for this publication is available from the British Library

Library of Congress Cataloguing in Publication data

ISBN 978-0-521-88085-5 hardback
ISBN 978-0-521-70708-4 paperback

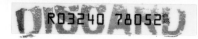

Contents

Figures

Preface

Chance has become an integral part of contemporary science but, for the most part, is still not at home in theology. Theology speaks of a purposeful God while chance, by very definition, seems to signify a total lack of purpose. To suggest the very opposite – that chance lies within the purposes of God – may seem perverse, if not foolhardy, and yet that is precisely what is argued for in this book.

One might have expected that the slow seepage of evolutionary theory or quantum theory into the public consciousness would at least have softened the hard edges of the confrontation. But – to change the metaphor – the Intelligent Design debate has fanned the dying embers into flames. 'Atheistic evolution' is now set in stark opposition to the theistic design which Intelligent Design sees as responsible for the wonder of the universe. This, therefore, is an opportune moment to argue again for the positive benefits which flow to theology from seeing chance as an intended part of the creation.

As is so often the case when matters are hotly disputed, the stark conflict between opposites begins to dissolve when we examine them carefully. Much depends on the 'level' at which we choose to observe the world. Many of the regularities of nature are built on randomness and much of the seeming randomness around us is like a shower of sparks thrown off by the lawfulness of processes on a larger scale. Self-organisation in apparently disorganised systems occurs on such a scale that one suspects it does not emerge by accident. It is just possible that primeval chaos was the precondition of a lawful world. This is an exciting question which is probed from several angles in this book.

I make no pretence that this is an easy book, even for those who are familiar with, if not expert in, the technical material which lies behind it. Many of the big recurring questions with which the theological giants have wrestled over the centuries reappear in these pages. The theology of chance is no magic solvent to dissolve the problems of yesteryear. What I do claim is that there is a great deal of misunderstanding which can be dispelled by rigorously subjecting widely accepted arguments to the critical eyes of someone trained in the statistical sciences. These are not, in themselves, intrinsically difficult so much as unfamiliar but the reader must expect to read slowly and to go over the ground several times, perhaps. I hope that the use of familiar, even homely, examples in place of mathematics will help. As an additional help, each chapter begins with a brief summary of its argument.

I suspect that chapter 7, refuting Dembski's claim to have produced a method for eliminating chance as an explanation, will be particularly challenging. It is, perhaps, also the most important chapter. Although I have reduced the technical matter to a minimum, it is essential to engage with Dembski on his own ground and in his own language. He has not spared his lay readers and I have, to some extent, followed his example.

There is, in fact, almost no technical discussion in the book which requires formulae for its expression and very little use is made of symbols. This does not really make things any easier because the economy and clarity of a mathematical argument makes it, typically, briefer and more precise than the equivalent verbal version. Its absence however, does make the text more reader-friendly and that counts for much. This observation should not lull one into thinking that probabilities, for example, can be plucked from thin air by some sort of innate intuition. Probability theory provides the backbone of reasoning about chance. Although it is not necessary to understand the theory to grasp the message of the book, it is important to follow the gist of the argument, and in particular, to be aware of the pitfalls which await the unwary. For this reason, some chapters represent a marking time in the progress of the main argument as I secure the foundations, albeit informally.

This is not the first time that I have ventured into this field. My *God of Chance* appeared in 1984. It is now out of print and seriously out-of-date although it is still available and may be downloaded free of charge at www.godofchance.com. Its primary aim, appropriate at the time, was to counter the implication of books such as Monod's *Chance and Necessity* that science had finally banished belief in God to the obscurantist backwoods. At that time I was not as well informed as I ought to have been about the late Arthur Peacocke's thinking in the same direction. The aim here is more ambitious, namely to give chance its rightful place in serious theology. My *Uncertain Belief*, first published in 1996, discussed reasoning about theological matters in an uncertain world. The uncertainty there was in our 'heads' rather than in the 'world'. For the rest, articles, lectures and chapters in books have provided the opportunity to develop many of the strands of thought which have come together in this book.

Chapter 14 is a shortened and updated version of the Gowland lecture which I gave to the *Christ and the Cosmos Initiative* at Westminster College, Oxford in 1999. It represents a first attempt to draw out, on a small front, the kind of theological thinking that is becoming imperative.

The footnotes serve the usual purpose of leading the reader into interesting and relevant byways. However, the reader with access to the Internet can track down a wealth of additional information by using a search engine such as Google. Occasionally web addresses are given here but the use of a search engine is so easy that they are almost superfluous.

As a 'solitary' thinker I may not acknowledge, as readily as I should, the debt I owe to others. Most of those from whom I have learnt much will recognise their contributions in these pages. I now find it impossible to identify and properly apportion credit where it is due. However, the Science and Religion community in the United Kingdom and internationally is a constant reminder of the interdependence of all thinking in this field.

In a book of this kind a disclaimer is essential. Anyone attempting to span so many disciplines must, inevitably, echo St Paul, who asked, in a different context: 'who is sufficient for these things?' (2 Cor. 2.15) The short answer, of course, is: no one. But if the apologetic task is not to go by default someone must take the risk. My only claim is that I am a statistical scientist whose work has taken him into many other fields. As one of the most famous of my colleagues (the late John Tukey) once said, 'the best thing about being a statistician is that you get to play in everyone else's back yard'.

It is becoming increasingly difficult to know how to thank my wife adequately for her continual and very practical support. Indeed, in these gender-sensitive days, it is hard to know whether these things should be mentioned in public at all! Whatever the proprieties, I cannot conclude without remarking that Proverbs 31.10–31 gets it about right.

CHAPTER I

What is the problem?

The problem is to reconcile the central place which chance has in the scientific account of the world with the theological account of God's relationship to the world. Chance suggests lack of purpose; theology speaks of purpose. This long-running source of tension has come to the fore again in the claims of the Intelligent Design movement, which aims to eliminate chance in favour of design. Quantum theory, which places chance at the heart of matter, poses essentially the same question for theologians. This chapter sets the scene and, very briefly, points the way to a solution which lies in seeing chance within, not outside, the providence of God.

CHANCE VERSUS GOD

Chance has become a major weapon of those who regard science and theology as locked in mortal combat. On the theological side there are those like Sproul,[1] who signals his intentions in the title of a book *Not a Chance* (1994). The subtitle makes

[1] Dr R. C. Sproul is an American theologian in the strict Calvinist tradition. He is a prolific author and the chairman of Ligonier Ministries, which he founded. The book, from which these quotations come, appears to be his only excursion into the science and religion field though his argument is primarily directed against what he perceives to be the faulty logic used by mainstream scientists. The quotations used here were chosen because they express, with great clarity, an extreme position adopted by some Christians.

his intentions doubly clear: *The Myth of Chance in Modern Science and Cosmology.* In the preface he goes on to say 'this book' may be viewed as a diatribe against chance. 'It is my purpose to show that it is logically impossible to ascribe any power to chance whatever.' As if that did not make his intentions clear enough he continues, on page 3, 'If chance exists in its frailest possible form, God is finished . . . If chance exists in any size, shape or form, God cannot exist.'

These are strong words indeed and one can only marvel that such an annihilation can be accomplished in hardly more than two hundred pages. Sproul is not alone, of course, though few other protagonists claim quite so much. Overman[2] is another who has entered the lists with his *A Case against Accident and Self-organization* (1997). This is a more sharply focussed attack and with more technical apparatus, but its intention is much the same. Those unfamiliar with the probability logic involved may be easily impressed when he concludes: 'The probability of chance causing the formation of a universe complete with life and the first forms of living matter is less than the mathematical impossibility at the accepted standard of 1 in 10^{50}' (p. 181). It is not clear where he acquired this 'accepted standard' or who accepts it! We shall return to the matter of very small probabilities later.

---

[2] Dean L. Overman is a distinguished lawyer based in Washington DC. The foreword to his book was contributed by Wolfhart Pannenberg and there is a commendation on the flyleaf by Alister McGrath. The laudatory remarks on the dust cover include quotations from Owen Gingerich and John Polkinghorne. It seems generally agreed that this is a detailed and clear approach to a very important topic, but endorsement of the conclusions reached is not so easy to find. In my judgement the central conclusion, like many of its kind, is based on a fallacious probability argument as will be shown later.

From the science side there are equally forceful advocates *for* chance. Although it was written more than thirty years ago, Jacques Monod's *Chance and Necessity* remains one of the most eloquent statements of the contrary view. Although often quoted, it bears repetition.

> We say that these events are accidental, due to chance. And since they constitute the *only* possible source of modifications in the genetic text, itself the *sole* repository of the organism's hereditary structures, it necessarily follows that chance *alone* is at the source of every innovation, of all creation in the biosphere. Pure chance, absolutely free but blind, at the very root of the stupendous edifice of evolution: this central concept of modern biology is no longer one among other possible or even conceivable hypotheses. It is today the *sole* conceivable hypothesis, the only one compatible with observed and tested fact. And nothing warrants the supposition (or the hope) that conceptions about this should, or ever could, be revised. (1970, p. 110; trans. Wainhouse 1972)

Oddly enough, both sides agree that chance eliminates God. Sproul's remedy is to oust chance; Monod's, to do the same to God.

The object of this book, roughly speaking, is to bridge the gap by saving Sproul's theology and Monod's science using chance as the link between them. Sproul was wrong in seeing chance as a threat to the sovereignty of God and Monod was wrong in seeing chance as eliminating God. These are strong claims and their justification rests, essentially, on the claim that chance must be seen as lying *within* the providence of God and not outside it.

As I pause to draw breath it is pertinent to remark that whatever chance is, it is certainly not an agent capable of *causing* anything, as Overman supposes. Even such a careful writer as Dowe (2005) falls into this trap when he writes 'If the latter is true God does not cause any event caused by chance'

(p. 184). But there is a broader field to survey before we return to this matter.

The essential problem is how to accommodate within a single world-view the element of real chance, which science seems to require, and the existence of a God who is supposed to be actively involved in creating and influencing what happens in the world.

A good deal clearly hangs on what we mean by *chance*. This is not such an easy matter as it may seem and it forms the subject of chapter 2. A chance event arises when something happens which we could not predict, but this may be because we do not have enough information. Chance is then the other side of the coin to our ignorance; this is sometimes called *epistemological chance*. Alternatively, chance may be *ontological*. That is, it is somehow inherent in the nature of things and there is no knowledge we could possibly have which would make any difference. This brings us to the crucial issue of God's involvement in the world because this depends on what view we take on the nature of chance.

Nevertheless there are some situations where we do not need to answer the deep philosophical questions. In particular, these arise when we come to *calculate* probabilities. The theory of probability is not so much about what probability *is* as about how to make probability calculations about uncertain happenings. Any attempt to calculate the probability that life would appear on earth, for example, depends upon putting together the probabilities of simpler, constituent events which were necessary for life to appear. This is where Overman, and many like him, have gone wrong.

Next there are rather crucial questions about what is implied by the existence of chance happenings in the world. It is commonly assumed to be self-evident that any intrusion of chance will lead to unpredictability and uncontrollability. This is not necessarily so and quite the opposite may be true. We shall discover that there may be extreme constraints which render the outcome of some chance processes almost certain. There is also an important matter of *levels*, to which we come in a moment, where what is uncertain at one level may be virtually certain at another level. Dying, for example, is still a highly unpredictable matter for individuals, but insurance companies and undertakers make a steady living out of it because what is individually uncertain is highly predictable in the aggregate.

There are equally weighty questions to be raised on the theological side about the nature of God. It was Sproul's determination to defend the sovereignty of God that led him to conclude that chance was impossible in God's world. Is it really true that absolute sovereignty requires that God knows and controls every one of the trillion upon trillions of events that occur in the universe every second? Might it not be that such a view actually diminishes the greatness of God? I shall, in fact, argue that this is the case.

More important, perhaps, is the effect which uncertainty in the world has on what God can know – his omniscience. Can he know, for example, what is as yet undetermined? Can we be truly free in a world in which God controls every single thing? I raise such questions now merely to show that the chance issue is not peripheral, but goes to the heart of age-old questions which have become even more pertinent in a scientific age than they were in the early centuries after Christ, or even in the Middle Ages.

New questions arise which can only be phrased in the language of chance or risk. We have to ask not merely whether God can cope with chance or even use it to good effect, but whether it might have a more positive role. After all, we use chance in large-scale computer modelling to mimic the uncertainty of the world, and to achieve goals which lie beyond our reach without it. If we find that risk taking can be beneficial and not always a necessary evil, may this not open new perspectives for theology? Can we conceive of God as a risk taker? This is, perhaps, the key theological question and we return to it in chapter 14.

Chance and providence go together in popular debate because each seems to be at variance with the other. How can God act providentially in a world if it is not wholly under his control? This question is distinct from, but not unrelated to, the question of whether and how God can act in the lawful world revealed by science. What kind of a place the world is also affects what we can know about it. If there is genuine uncertainty about what we observe, then presumably there is some uncertainty about what we can infer about it. More especially, and challenging perhaps, is the question of what we can know about God and his purposes for us and the whole creation. Are there any certainties left for the believer or, indeed, the unbeliever?

Next, there is the relationship between chance and law. The two seem to be in direct opposition but this is not necessarily true. In fact some laws have been correctly described as statistical, or probabilistic. These are laws which relate to large aggregates and thus operate at what I have called a different level. The simplest possible example is the tossing of a coin. The outcome of a single toss is a highly uncertain matter but the outcome of 10 million tosses is highly predictable in the sense that we can say that almost exactly 50 per cent of all tosses will be heads and, furthermore, we can also be precise about what deviation from that average figure is likely to be. The constancy of such ratios certainly has the law-like characteristic we expect in dealing with a system of divine origin and concern. The gas laws are a more interesting example, where the relationship between the pressure and volume of a gas at a constant temperature is simply determined by the average effect of a very large number of gas molecules. Lawfulness at the higher level of aggregation is thus the direct consequence of complete randomness at the lower level.

This intimate relationship between levels of aggregation also works the other way round: lawfulness can give rise to chaos. The word chaos is used here in its technical sense but, for present purposes we can think of it as, more or less, equivalent to chance. Very simple, law-abiding processes can give rise to chance. The subtlety of these relationships emphasises that what we see in the world depends on the level at which we enter it to make our observation. The same is true in the realm of physics where, as we shall see, the quantum world seems very different to the everyday world viewed on the human scale.

As we unravel the complexities of these relationships it will become apparent that Sproul's and Overman's categorical statements are not so much wrong as inapplicable to the world in which we actually live, or to the God who created and sustains it.

LEVELS AND SCALES

Before we go any further I must digress to expand on something which has already cropped up several times and which is central to the question of God's sovereignty.[3] What we observe in the world, and how we describe it, depends upon how big

---

[3] The idea of levels and scales occurs in other contexts. A recent example is provided by *The View from the Centre of the Universe* by Primak and Abrams (2006). The prime object of that book is to argue that, in a certain sense, we are at the centre of things and that gives us significance as humans. For example, humans are at the centre of the scale of size. If length is measured in orders of magnitude (that is powers of ten) then lengths on the human scale come somewhere near the middle of the range which extends from the smallest things we know (the Planck length of 10^{-33} cm) to the

or small a scale we view it on. The world viewed through a microscope is very different from what we see through a telescope. On the astronomical scale we simply do not notice the biological details of nature. The microscopic world has no place for mountains and trees. They are too big to be viewed on such a small scale.

If we go to the limits of size, in either direction, the worlds we find are beyond our imagining and the best we can do is to describe them by mathematical equations. On the very large scale, the Euclidian geometry of the schoolroom lets us down and we then have to reckon with things moving close to the speed of light. Space has to be thought of as curved and we find ourselves in a world in which Newton's physics is inadequate and our imagination fails.

Something similar occurs at the smaller end of the scale where we get down to the level of atoms and what goes on inside them. It is an unfamiliar world in which our intuitions based on the everyday world simply do not work. The mathematics continues to work perfectly and delivers results which are entirely consistent with the world as we perceive it at our level, but any attempt to picture it fails.

For practical purposes, we can think of three levels. First, the everyday world of things that we can see, touch and handle, where distances are measured in metres or miles – but not in

largest (the distance of the cosmic horizon of 10^{28} cm). Similarly, if less convincingly, human life has occurred in the middle of time measured from the beginning to end of the universe or the life of the earth. The important thing, from the present perspective, is that according to the authors, certain questions only have meaning – and hence meaningful answers – if posed at the appropriate level. This idea is applied in many fields including the nature of God. 'God' must therefore mean something different on different size-scales yet encompass all of them. For example, all-loving, all-knowing, all-everything-else-we-humans-do-only-partially-well may suggest God-possibilities on the human size-scale, but what about all the other scales? What might God mean on the galactic scale, or the atomic?

light years – where weight is measured in pounds and tonnes, and so forth. This I shall often refer to as the *human level*, or scale.

Secondly there is the world of the very small – too small to be observed by the naked eye – the world of cells, molecules, atoms and electrons. At best we can only see parts of this world through microscopes but often what is going on has to be observed indirectly through the observable consequences of what is happening at the micro level beyond the limits of our direct observation. This may conveniently be designated the *micro level*.

Finally, there is the world of the very large where, for example, masses can be so large as to produce observable deflection of a beam of light. In this world we encounter incomprehensibly large numbers and unimaginably long periods of time. This I shall call the *macro* or *cosmic level*. We can sometimes be helped to grasp the significance of these things if they are scaled down to something we can understand. For example, the relative distances between the planets in the solar system can be represented by where they would appear if they were laid out on a football pitch with the distances between them in the same proportions.

The concept of scale or level is central to understanding the place that chance occupies in the grand scheme of things. For example, what appears chaotic at one level may reveal a pattern when viewed on a larger scale.

This phenomenon is familiar to computer users through the zoom-in and zoom-out facility which many computers offer. As we zoom in we see more and more of the detail and less and less of the overall picture. Conversely, when we zoom out the reverse is true. This is very obvious when viewing a map. At the lower level we see individual streets, whereas at the higher level these merge into a blur as the shape and location of the town becomes the dominant feature. When viewing text at large magnifications we lose sight of the words and notice only the patchiness of the individual letters. When we take the broader view, the words disappear and we begin to see the pattern of the layout and so on. Each view shows a different aspect of reality.

In our world it is natural to think in terms of the human scale on which we live our lives and form our understandings and intuitions. It is in this world that we form our concepts. It is in this world that Christians believe God revealed himself on a human scale. The truth thus revealed makes sense to us because it is on *our* scale. It does not follow, of course, that God's actions can be necessarily or exclusively understood at our level. In fact, as we shall see later, there have been valiant attempts to account for God's providential action by reference to happenings at the micro level.

Since, presumably, God and his creation are not commensurable we must be very wary of creating a God in our own image and on our own scale. The intimate connection between chance and order at different levels of the creation, which has been noted above and which I shall explore later, make it very important to be careful about how we use language. To suppose arbitrarily that God's main sphere of action is at the level we can most easily comprehend may be a dangerous and misleading assumption. This is relevant to two issues which are of great interest in themselves and which have brought chance to the fore in contemporary debates.

INTELLIGENT DESIGN

One of the most extraordinary phenomena to have arisen on the science and religion scene in the last few decades is the Intelligent Design[4] movement. This is largely an American

---

[4] There is an immense literature on Intelligent Design. In this present book we are concerned only with the logic of the argument by which, it is claimed,

phenomenon and it is fed by the peculiar mix of fundamen-
talisms which flourish there. In a sense it springs from the
acute concerns of those, such as Sproul, who fear that chance
strikes at the root of Christian belief. The great enemy, as its
protagonists see it, is the naturalism of modern science. This
refers to its attempt to explain everything that happens without recourse to any external direction such as is traditionally
supplied by God. Chance and necessity, in Monod's mem-
orable phrase, account for everything. Even well-meaning
Christians, who see evolution as God's way of creating things,
are deluding themselves, opponents would argue, and they are
embarked on a path which will inevitably leave no room for
God. Methodological naturalism is also viewed with great
suspicion. This is the strategy of proceeding *as if* everything
could be explained without reference to any external creator
or designer. For if you really believed that there is clear evidence of design in the world it would be foolish to ignore it,
as a matter of policy, and so have to compete with one hand
tied behind your back!

The search for evidence of Intelligent Design, which I shall
examine in some detail in chapter 7, involves the attempt to
eliminate chance. If this could be done, design would remain as
the only, and obvious, explanation. Intelligent Design assumes

---

chance can be eliminated as an explanation of evolutionary development.
William Dembski has developed this single handedly and his work, therefore, is the focus of our attention. A fuller account would have to take note of
the work of Michael Behe, especially his *Darwin's Black Box: the Biochemical
Challenge to Evolution* (1996). Another key figure is Phillip E. Johnson, also
a lawyer. It is pertinent to note that lawyers approach things in a rather
different way to scientists. Lawyers operate in an adversarial context where
their object is to detect and expose the weaknesses in the opponent's case.
Science is an ongoing activity in which the well established and the provi-
sional often exist side by side. It is not too difficult to find weak points in
any scientific theory. This is a valuable thing to do but it does not establish
what is true or false in any absolute sense.

at the outset that chance and design cannot coexist and this claim is totally contrary to the thesis advanced in this book. If the Intelligent Design movement were to be successful, the ideas to be set out here would be completely undermined. It is therefore essential to examine the logic of the arguments of those, such as William Dembski, whose highly technical treatment of inference under uncertainty underpins the whole enterprise.

At this and other points on our journey it will therefore be necessary to examine the logic of inference under uncertainty. It is to Dembski's credit that he recognises the need to provide a rigorous account of what is needed to eliminate chance as an explanation of any phenomenon. He also recognises that much of the groundwork has already been done by statisticians and he cites Sir Ronald Fisher[5] as one of those who have paved the way for his own new developments. Unfortunately for him, Dembski's ambitions founder not only on the faulty logic of his inference procedure but on the calculations necessary to implement it.

DOES GOD ACT IN THE WORLD?

When we look at the world on the very small scale there is a whole new territory on which to debate the role of chance. Oddly, this field does not appear to have had any serious inter-action with the evolutionary issues which have so exercised

---

[5] Sir Ronald Fisher was, perhaps, the leading statistician of the twentieth century, though some of his ideas, including those on significance testing, were controversial and are not widely accepted today in their entirety. He was, successively, Galton professor of Eugenics at University College London and Arthur Balfour professor of Genetics in the University of Cambridge. A full account of his life will be found in his daughter's biography: *R. A. Fisher, the Life of a Scientist* (Box 1978).

the debaters of Intelligent Design. Classical mechanics has been very successful in describing the dynamics of the world of everyday objects such as tables and tennis balls but when we look at things on the scale of electrons and photons it breaks down. In that case we need quantum mechanics, which was specially developed for work at that level. Using quantum theory it is possible to be quite precise about calculations made relating to quantum phenomena. The trouble is that the theory provides an incomplete description of reality at that level. This is of no great inconvenience to physicists, who can make their calculations regardless, but it poses serious questions for philosophers and theologians. The latter, especially, want to know what is really going on, so that they can examine whether it is an appropriate arena for the action of God. There are competing interpretations of what the quantum world is actually like, and it is those that incorporate an element of chance that fall within our present concerns.

At one extreme are those who prefer a wholly deterministic interpretation of the quantum world. This is possible, if somewhat contrived, and it certainly fits in with the worldview of theologians such as Sproul. At the other extreme, what is prescribed are probabilities. According to that view, the individual events can then occur as God wills provided that, in aggregate, they conform to the overall probabilities. One way of describing this situation is to say that we only know the quantum world through probability distributions. This enables us to predict where particles are and what they are doing probabilistically but these distributions do not provide definitive information about what is actually the case. So chance is seen as a positive asset because it provides room for manoeuvre for a God intent on purposeful activity. Chance is then not the enemy of theism but necessary for it to be credible. The problem posed by the possibility of God's acting in quantum events is not essentially different from his action in relation to any statistical law. This will be a recurring theme in the following chapters.

GOD'S CHANCE

In chapter 11 we come to the heart of the matter and I argue for the view that chance is a deliberate part of God's creation. Not only is the presence of chance an integral part of the created order but it actually offers possibilities of variety, flexibility and interest which would not be available in a deterministic universe. This contention does not, of course, sweep all opposition before it. There remains a cogent case to be made in defence of a more traditional theology. However, this is not to be found by following Sproul, Overman and others down the road of anti-science or of pseudo-science. It is more likely to be found in arguments such as those of Byl (2003), whose case I answer in chapter 12. But there is also a paradox in introducing chance as a way of providing freedom of choice and then having to reckon with what this does for the rationality of human choosing. This is the subject of chapter 13. In the final chapter I deal at greater length with what I see as one of the most serious theological challenges posed by the God who works, in part at least, through chance.

CHANGING PERSPECTIVES

One can detect a progression as one moves through the debates which provide a connecting thread throughout this book. First there is the idea that chance is a problem for theology. Its perceived existence seems to challenge the sovereignty of God and call for scientific effort to be devoted to its elimination. Only when this is done will the true nature of reality be revealed.

This was the thrust of much early work designed to show that the complexity and wonder of the world revealed by science simply could not be 'due to chance'. Overman cites many examples. Some of this work is reviewed in chapter 6 but this line of thinking finds its culmination in Dembski's approach to Intelligent Design treated in chapter 7.

In the second stage chance has a more benign role. It is seen as playing an essential, but passive, part in providing the space for God to act without disturbing the lawfulness of the world. The trouble with the fully deterministic system, which a rigorous view of God's sovereignty seems to require, is that it leaves no room for free action either on God's part or our own. If we can create space for free actions, without disturbing the order in nature or requiring God to contradict himself, then progress will have been made. Quantum theory, according to some interpretations, appears to allow just the flexibility that is called for.

More generally, the widespread occurrence of statistical laws in nature and society seems to provide further room for manoeuvre for both God and ourselves.

Neither of these two approaches is satisfactory and a substantial part of this book is devoted to exposing their weaknesses and preparing the ground for what I believe is a more adequate view. This sees chance in a more positive light, as something which actually does greater justice to the sovereignty of God and to his remarkable creativity. Freed from an excessively 'mechanical' way of thinking about God's actions, we see an enormously rich tapestry of opportunities and possibilities in the creative process. In short, chance is to be seen as within the providence of God rather than outside it. It is a real part of the creation and not the embarrassing illusion which much contemporary theology makes it out to be.

CHAPTER 2

What is chance?

This chapter aims to clarify the terminology. In a field where there are many terms with overlapping meanings, it is important to establish a common language. The approach is through the familiar notion of uncertainty, and the chapter continues by dealing with chance in relation to ignorance, accident, contingency, causation and necessity. Particular attention is paid to the idea of 'pure' chance, and the paradoxical relationship of choice and chance is touched upon.[1]

WHY THE QUESTION MATTERS

This is a fundamental question for theology, especially if we are to understand why some think that the very idea excludes God. Sometimes chance is spelt with a capital C as though to accord

[1] Chance is commonly set in opposition to purpose as in the title of this book. It is therefore important to raise at an early stage the question of what it means. This is not a straightforward matter, because it is used with so many variations of meaning, especially when we come to philosophy and theology. Originally, as in Abram de Moivre's famous *Doctrine of Chances*, the first edition of which was published in 1718, it had the purely technical purpose of referring to chance events such as rolling dice or drawing cards. This use persists in speaking of games of chance and in much of scientific writing. However, chance has more emotive connotations as in 'blind' chance, and this usage has become a common weapon in the debates on evolution versus creationism. The prime purpose of this chapter is to draw attention to the subtleties of the situation and to prepare the ground for the discussion of probability in chapter 5. This provides a more secure foundation for what follows.

16

it a quasi-metaphysical status. I have already noted, in chapter 1, the hazards of thus personifying chance because to do so almost inevitably sets it up in opposition to God. It is then but a short step to speaking of chance as causing things, and once that happens we are on a very slippery slope indeed. The fact that we have so many words for chance with an overlapping meaning in our vocabulary shows the subtlety of the concept. Random, likelihood, probability, contingency and uncertainty are all terms which arise and sometimes they are used in conjunction as when people speak of 'random chance' as though to emphasise the sheer uncertainty of it all! In the following section I seek to elucidate the meaning of *chance* by looking at it in the context of some other terms with which it is commonly linked.[2]

CHANCE AND UNCERTAINTY

Chance and uncertainty are closely related. In fact I shall conduct the present discussion, initially at least, in terms of uncertainty, because this word seems less laden with metaphysical overtones than *chance* and thus has the merit of being something we would never think of personifying. Life is undeniably uncertain; we confront uncertainty at every turn, so this provides a gentle, if somewhat roundabout, route to our final goal. What the term lacks in precision is made up for by its familiarity.

---

[2] The late Arthur Peacocke wrote extensively on chance and its role in the world. Much that is relevant to this chapter, and the book as a whole, will be found in his chapter entitled 'Chance and law in reversible thermodynamics, theoretical biology and theology' in Russell *et al.* (1995). See especially the sections on Chance, Two meanings of chance, God and chance, and Living with chance.

To say that some event happens *by chance* is to say no more than that we do not know enough about its antecedents to predict its outcome with certainty. This leaves open the question of whether our failure to predict is due simply to lack of knowledge on our part, or whether there is no knowledge available, even in principle, which could give us a degree of certainty. In the first instance all uncertainty arises through lack of knowledge.

We are uncertain about many things and events because our knowledge is limited. Certainty is where we arrive when we have all the knowledge that is relevant, so it is a good idea to begin with certainty as the end point of the scale of uncertainty. We are certain that the sun will rise tomorrow because we know that 'tomorrow' is defined by the rotation of the earth and its position in relation to the sun. As the earth revolves, the sun repeatedly comes into view and then sets again. Philosophers, of course, might dispute the certainty of this, and distinguished members of that fraternity have questioned, in all seriousness, whether it really is certain. Here, however, I do not seek to set up a watertight argument, but to conduct an enquiry at a commonsense level. Death and taxes are the traditional certainties but there are many others. If we drop a lead ball from an unobstructed church tower, we are certain that it will hit the ground. If we are bitten by a poisonous snake, we are certain to be very unwell. If we refuse all food and drink for a sufficiently long time, we shall certainly die.

Actually, very few happenings are as certain as that. If we are exposed to someone with a cold, it is not certain that we, ourselves, will develop one soon afterwards. If we set out on a train journey, it is uncertain that we shall arrive on time. If we plant a hundred pea seeds, the number that will germinate is uncertain. The difference between the certain events and the uncertain events is in the degree of knowledge we have in each case. In the 'certain' cases we know all about the circumstances and how the 'mechanism' determining the outcome works; in the uncertain cases our knowledge is incomplete.

There are degrees of uncertainty. We may know some relevant things but not enough to be sure about what is going to happen. The weather, especially in countries such as Britain, is a highly uncertain affair. But weather forecasters know a great deal about what determines the weather and, though they may not always be right in their forecasts, they usually do a better job than untutored amateurs. Weather forecasters do better than we can because they have more relevant information than we do and are able to handle it more efficiently. As more knowledge about the atmosphere has become available so, by and large, the uncertainty in their forecasts has decreased. Thus there is an inverse relationship between uncertainty and relevant knowledge. Generally speaking, uncertainty decreases as the accumulation of relevant knowledge increases. The general drift of this argument so far suggests that if only we could acquire enough knowledge, we would be able to banish uncertainty. Whether or not this is so is actually an interesting and fundamental question. Is it, in fact, true, I ask again, that uncertainty is solely the result of removable ignorance? Or is it possible that some events are unpredictable in principle – at least to some degree? This would mean that however much knowledge we had, some things would remain irreducibly uncertain.

Statisticians often behave as if there were an irreducible uncertainty about the things they try to predict. When trying to predict the sales of, say, a new computer they will typically identify those variables that seem to be important (e.g. number already sold, performance of the new model, etc.) and put these into their equations. (It does not matter for present purposes where these equations come from or what form they take.)

These will not be sufficient to give perfect predictability, so statisticians will then add what is variously called an *error term* or a *residual* whose purpose is to capture the effect of everything else that has been left out. It is a sort of balancing item because it is what it would be necessary to add to make the sales fully predictable. In practice, of course, they will never know the value of this balancing item but there are sometimes ways of estimating it if we have replications of the data.

In certain very simple situations one can come very close to perfect predictability with very few predictors, but this is the exception rather than the rule. In most cases, especially in the social sciences, the predictability is poor even after the list of known potential predictors has been exhausted. The question then arises as to whether, even in principle, it would be possible to achieve perfect prediction if one had complete knowledge. If not, then there would seem to be an irreducible degree of unpredictability. This, in turn, touches on deep questions of what, if anything, determines outcomes in the absence of any causal factor. For the moment, we must be content to leave that question unanswered.

CHANCE AND IGNORANCE

It is now time to move away from the focus on uncertainty and link uncertainty more specifically with the technical terms of chance and probability. Both are clearly related to uncertainty through the fact that there are degrees of uncertainty. But uncertainty and probability are not identical and part of the price I must pay for the opening digression through the more familiar territory of uncertainty is a partial retracing of my steps.

We now return to the subject of epistemological chance mentioned earlier. Even if everything were perfectly predictable it does not follow that the study of uncertainty would be redundant. Complete knowledge may be *practically* unobtainable, so the world appears much as it would if it were unpredictable in principle. In other words, we may be ignorant. There are a great many situations in which we know that there is further information available, at a price, which would improve our predictions – weather forecasting is a case in point. In the social sciences this situation is the norm. In practice, therefore, we are usually working with chance as reflecting a degree of ignorance. These degrees of ignorance, or knowledge for that matter, are measured by probability, to which we come in chapter 5 but, for the moment, I note that the calculus of probability takes no cognisance of how the probabilities with which it works are defined. Uncertainties, whether they are irreducible or simply descriptions of our ignorance, are all grist to its mill. Probability, as we shall see, is a measure of uncertainty which conventionally ranges between zero and one. A value of one denotes certainty and zero indicates impossibility. Uncertainty diminishes as we approach *either* extreme. For example, we may be almost certain that our friend will keep a promise to keep an appointment with us. This amounts to saying that there is a high probability of the appointment being kept. Equally, we may be practically certain that it will not snow on Midsummer's Day, for which the probability is almost zero. Very high *and* very low probabilities thus correspond to low uncertainty and vice versa.

CHANCE AND ACCIDENT

Accidents happen when two or more causal chains coincide. An accident is, therefore, a coincidence. Coincidences are not usually predictable, by us at any rate, because we do not have sufficient knowledge to foresee what is certainly going to happen. An all-seeing and all-knowing being, like God, would be able to see the whole situation as it developed and for him there would be no chance involved. Chance and accident could, therefore, have been subsumed under the general heading of 'Chance and ignorance' but accidents involve the lack of a special kind of knowledge, namely about the antecedent causal factors. It is in the apparent independence of the causal chains that the source of the uncertainty lies.

CHANCE AND CONTINGENCY

Contingency is a philosophical term referring to the fact that some happenings are not determined – that things could have turned out otherwise. It is sometimes used as a synonym for uncertainty, accident or chance, but in the present context it is liable to be ambiguous. It might mean determined by God or accidental in the sense defined in the last section and that distinction can be very important. In her book *Divine Will and the Mechanical Philosophy*, Osler (1994) explores the attempts of philosophers and theologians of the seventeenth century to get to grips with the way that God related to the created world. The subtitle, *Gassendi and Descartes on Contingency and Necessity in the Created World*, shows us that their efforts crystallised around contingency and necessity. This resonates with the *chance and necessity* which has come to epitomise the essence of evolutionary theory, but the similarity of language conceals a major difference. Both sides in the earlier debate accepted that God was responsible for everything. The point at issue was whether some things happened of necessity because God had planned the world to work that way or whether, as Pierre Gassendi thought, God had complete freedom to act as he willed on every occasion. Chance (and fortune) in their understanding was more in the nature of accidents which are only uncertain to us because we do not have full information. As we shall see, contingency in that earlier sense is still an option canvassed by many theologians but it is radically different, in the view of most scientists, from the chance that drives evolution.

PURE CHANCE

The *pure* in pure chance suggests a lack of contamination which is a fair way of describing the absence of any predictive factors. Pure chance is what we get when there is nothing at all which has any predictive value. We can, therefore, never be sure that a chance event is pure, in that sense, because we can never be sure that there is not some elusive and unobserved factor which might help us to predict and so remove some of the uncertainty. The archetypal example of pure chance appears to be provided by radioactive decay. The word 'appears' is crucial in this sentence because we can never know for certain that there is no possible predictor lurking somewhere in the universe that we have failed to notice, but it is the best example we have. Nothing is known which helps us to predict when the next emission from a radioactive source will occur. The term *pure chance* is sometimes used as a synonym for *ontological* chance, which distinguishes it from *epistemological* chance.

Must every event have a cause? If so, we have to find some causal agent for those happenings where pure chance has been invoked; otherwise we must abandon the idea that every event must have a cause. Paradoxical though it may sound, such events must be caused by *whatever*, or *whoever*, it is that causes apparently uncaused events! It is at this juncture that theologians have often felt they can save causality by introducing God as the sufficient cause of anything lacking a detectable cause in the physical universe. If we seek to avoid this dilemma by denying that all events have causes, we have to allow that there are some things that simply cannot be explained and leave it at that.

CHANCE AND CAUSATION

As just noted, the existence of chance events raises fundamental questions about causation. In the last chapter I introduced the idea of levels of reality. Appealing to the notion, made familiar to computer enthusiasts, of *zooming in* and *zooming out*, I noted that descriptions of the world at one level may appear quite different to those at another level. This applies especially to what I have said about uncertainty. We may be uncertain about a particular outcome of some event but have little uncertainty, at a higher level, relating to a larger aggregate of which the single event forms a part. The drawing, at random, of a card from a well-shuffled standard pack is a highly uncertain event and is, as near as makes no difference, a purely chance happening. But if we draw, with replacement a thousand times, the number of hearts will be very close to 250, or 25 per cent. There is thus no inconsistency in having a great uncertainty at one level and near certainty at the next higher level. It is, therefore, important to be specific about the level we are describing.

In order to give a firmer grasp of these important ideas it will be useful to relate them to a number of familiar situations in which uncertainty arises and which are germane to the present discussion. This I do in the next chapter.

CHANCE AND NECESSITY

In the titles of the other sections of this chapter so far chance has been paired with one of its synonyms to illustrate the diversity of usage and meaning. It may seem perverse, therefore, to pair it with its opposite. However, chance and necessity have become indissolubly linked by Jacques Monod's use of them as the title of his ground-breaking book. From that point onwards they have become part of the currency of the debate between evolutionists and creationists, usually in the form of chance versus necessity. In effect this has implicitly defined chance as that part of the evolutionary account which makes it anathema to anti-evolutionists.

But the chance referred to here is not the pure chance just defined but is much closer to accident. It arises in the evolutionary story where mistakes occur in the copying of the DNA. In practice the copying process is extremely accurate but very occasionally errors occur for a variety of reasons. When such changes occur they are described as mutations. Cosmic rays or other damaging effects, such as chemicals, can introduce such errors. There are causes for these errors which can be explained in physical terms, so it cannot be said that they are without cause. The justification for using the word 'chance' is that their effects are quite independent of the resulting change in the phenotype. In practice this lack of linkage makes it impossible to predict the consequences of the mutations. Mutations introduce uncertainty, and uncertainty spells chance.

Because of the fervour with which the role of chance, in this sense, is condemned as *blind* chance and therefore as anti-purpose, it may help to introduce an illustration. Children play a game known as Chinese Whispers. A number of children sit in a row. The child at one end whispers something to their immediate neighbour. The neighbour repeats what they think they have heard to their neighbour, and so on down the line until the message reaches the other end. In most cases the message will have become so distorted in the course of its being passed from one to another that the final result will cause much hilarity. It is difficult to obtain a better example than the, no doubt, apocryphal story (from pre-decimalisation days!) of the British general who sent the following message to headquarters: 'Send reinforcements; we're going to advance.' After it had passed through several hands HQ received it as: 'Send three and four pence; we're going to a dance.' The amusement of the game results from copying errors, or mutations. Each person in the line attempts to copy the sequence of sounds received from their neighbour and errors will inevitably occur. The analogy is not exact, of course, and one should not press it too far, but the point is that any errors should be independent of *meaning*. Changes of meaning are therefore, not predictable and the outcomes may be legitimately described as chance effects. Necessity is thus the opposite of chance because the former is precisely predictable and the latter is not.

FREE WILL

There is another realm where the existence of causal links is in serious question: the field of human choice. Are all our choices predictable? This would be so if, having observed enough things about the person's brain, their immediate environment and the universe in general, we could say with certainty what choice that person would make. Thus would free will be banished. Note that it is not enough to say that we can predict a decision in *some* situations, for there might be others where we could not, and that would save free will in some circumstances at least. So whether or not free will exists depends on whether our decisions and actions are always, in principle at least, predictable. There is then no uncertainty about the outcome of what goes on in the process of making up our minds. In practice, of course, we are likely to be in the 'in-between', state where we can observe things which reduce the uncertainty but do not eliminate it.

If free will exists in this sense, it must be impossible to obtain enough knowledge to predict, with certainty, what we shall do in every circumstance.

The definition of free will that I have just given bears a remarkable similarity to that proposed for pure chance. How something which seems to be the very epitome of rationality and purpose can be so similar to its antithesis is an intriguing question which will have to wait for a fuller treatment.

For a second time then, we see that asking simple questions about uncertainty touches on the most fundamental issues which have exercised philosophers and theologians down the centuries. The free-will versus determinism debate brings choice and chance together and in doing so creates a paradox which I shall not attempt to resolve until chapter 13.

CHAPTER 3

Order out of chaos

Much of the order in the world is built on disorder. This fact makes it difficult to speak unambiguously about whether or not it expresses purpose. For when many chance events are aggregated, order often appears. This chapter substantiates the claim that haphazard happenings at one level may lead to lawfulness at a higher level of aggregation. It begins with simple processes such as sex determination and moves on to the regularities which appear in networks of many kinds.

ORDER BY AGGREGATION

Much of the debate about whether chance and God are compatible centres around the alleged inconsistency of believing that God's purposes are constant and in recognising the uncertain behaviour which characterises much of the world we live in. The former speaks of God's presence and the latter of his absence. We thus seem to be presented with the stark choice: God or chance. Sproul, and those who think like him, cannot conceive of a world in which the two could coexist. The purpose of this chapter is to show that things are not as simple as this crude dichotomy suggests. Order and disorder are closely connected and one may be a precondition of the other. This and the following chapter explore this relationship and so prepare the ground for a positive role for chance.

The essence of the point I wish to make in this chapter is contained in the simple example of coin tossing that was used in chapter 1. A single toss of a coin is a highly uncertain matter yet the collective tossing of a million coins is a highly predictable event. Chance at the level of the single toss is replaced by near certainty at the aggregate level. The way the world looks and the randomness, or otherwise, of its processes depend very much on the level at which we choose to observe it.

The idea is not confined, of course, to the artificial world of coin tossing but is all-pervasive. The motion of a single molecule inside an inflated balloon is chaotic as it collides with other molecules but the combined effect of the motion of the billions of molecules inside the balloon is highly regular, maintaining the shape over a long period. This idea extends to the stability of the whole universe. At the lowest level, quantum theory describes what is going on in probabilistic terms, but at the macro level, at which we observe it all, much of that randomness merges into the lawfulness on which our science and engineering are built. For that reason laws are sometimes described as *statistical* and it has even been speculated that all laws may have that character.

Lawfulness is claimed to be a key characteristic of our universe. The stars remain in their courses and cannonballs continue to fall from high towers with a constant acceleration. Because we can rely on the way nature behaves, science is possible and so, in the more mundane affairs of everyday life, we can go about our business in a rational way. This all speaks, theologians tell us, of the faithfulness and constancy of God. It is all part of the doctrine of creation, but it should already be clear that it is a very partial view showing us what the universe looks like from our particular vantage point. For a God, who has other vantage points, the description may not be so simple.

There are different ways of regarding laws. The laws of nature can be viewed as descriptive, not prescriptive. According to this view, they do not tell us what *must* happen but what *does*, as a matter of fact, happen. Their very constancy then warns us of what we can expect to happen when, for example, we jump off a cliff, but they do not tell us that this must necessarily happen. Another way of looking at laws is as constraints imposed on what can happen. Water has a different chemical composition to wine and the conversion of the former to the latter is not in line with how the world normally works. But does it have to be like that?

Our view on these matters affects how we think about other aspects of Christian belief. For example, it does create some problems, especially in relation to miracles, as I have just implied in referring to water and wine. When we move into this realm there is often a subtle change in the way we regard laws. We begin to think of them as similar to laws enacted by parliaments or the governing bodies of sports. These prescribe how the governing bodies wish people, or players, to conduct themselves. Incentives and sanctions may be introduced to induce the desired behaviour, but they in no way *determine* the choices of citizens or players. Such laws can be broken and, in spite of the sanctions or incentives, they often are. If we think of God's laws in this way, instead of as saying what, necessarily, must be so, the whole discussion of God's action moves into a different realm. These preliminary remarks are intended in a cautionary sense to alert us to the need to be careful about how we regard statistical laws and the inadvisability of jumping to premature conclusions.

The introduction of chance into our world-view produces a radical change for both science and theology. Lawfulness of a certain kind arises from randomness and, indeed, is built upon it. This requires a reassessment of what such laws tell us about God, and how he might act in the world.

In this chapter we shall meet some examples of how order may arise out of chaos[1] and that will provide the backdrop for a reconsideration of God's relation to the world and his inter-action with it. To make the discussion clearer it will be helpful to designate the two levels with which we are concerned as the *individual* level and the *collective* or *aggregate* level. These are to be thought of as relative levels rather than the absolute levels for which I used the designations micro, human and macro earlier.

THE HUMAN SEX RATIO

A female child receives an X chromosome from each parent; a male child receives an X chromosome from the mother and a Y chromosome from the father. At birth there are about 105 males born to every 100 females. The matching of chromo-somes at any one conception is quite independent of any other mating – whether between the same or different parents. This corresponds almost exactly to what happens when we toss coins, where heads might represent male and tails represent female. The only slight difference is between the proportion of heads and the proportion of males. What I have to say is true for sex determination and for coin tossing, but the sex determination example is more interesting.

The outcome of any particular birth, male or female, is completely unpredictable – in the absence of prenatal scans or other checks. In that sense births might be described as chance events. The individual birth is an event at the individual level. On the assumptions I have made there is nothing we can observe prior to the event which can help us to predict the

---

[1] The word chaos has an everyday meaning which is virtually synonymous with disorder. It is in that sense that it is used in this chapter. It also has a technical meaning, as in mathematical chaos, to which we come in the next chapter.

outcome. However, if we observe a large number of births in a town over the space of a year, for example, the proportion of males will be very close to 51.2 per cent [(105 / 205) × 100 per cent] and this is something we can confidently predict. In the second case we have an example of a collective event and it exhibits a lawful character – something which occurs always and everywhere. It is a law of human society that the sex ratio is about 51 per cent. At the aggregate level it is almost the same as if alternate births were female. But at the level of the individual family of two or three children, the lack of variation in the latter case would have implications, of course!

We have moved a little beyond artificial things like coin tossing, to what is, perhaps, the simplest example that shows how a particular kind of effect can be both chaotic and lawful at the same time. Which type of behaviour we observe depends upon the level at which we choose to make our observation.

Lawfulness emerges from chaos in all sorts of interesting ways and I now supplement this simple example about sex determination with others which are not quite so simple. Together they provide the necessary backdrop to the discussion of how God might act in the world. Also, and paradoxically it might seem, this movement between the individual and collective levels is a two-way process. Chaos can emerge from order as well as order from chaos. This makes it very difficult to talk about the 'lawfulness' and 'unlawfulness' in absolute and unqualified terms.

POISSON'S LAWS

Siméon-Denis Poisson (1781–1840) had humble origins but became a leading figure in the French scientific establishment. He lived in turbulent times and it was said of him that he was one 'whose political flexibility was as remarkable as the rigidity of his scientific convictions'.[2] His original researches ranged widely, well beyond probability theory, to which he was a considerable contributor. It is ironic, therefore, that the law, or distribution, which bears his name, should have been a fairly insignificant part of his work – and not entirely original. It was known to Abram de Moivre more than a century earlier and has become part of every statistics student's introduction to the law through the efforts of Ladislaus von Bortkiewicz a century later. It was von Bortkiewicz who noted that the deaths from horse kicks in the Prussian army varied in precisely the manner required by the law.

For present purposes the Poisson process, as we may call it, provides the perfect example of how the notions of randomness and law are so intertwined as to reduce much of the theological debate to nonsense. To see this we shall take a step back and then bypass much of the historical and contemporary discussion of this topic.

Suppose we begin by asking what it means to say that events are occurring 'completely at random' in time. This is, surely, a fundamental notion if we are to think about how God might be involved in such happenings. There are plenty of things, even within our own experience, on which we can draw to get to grips with the concept of randomness in time.

Standing at a bus stop, especially in a large city, often seems to prompt the thought that the arrival of buses is random. The timetable may say that they should arrive at regular intervals but the vagaries of traffic and other hazards produce marked irregularities. This experience indicates a way in which we might begin to characterise randomness in time. If, on arrival

---

[2] This quotation comes from Bernard Bru's biography in *Statisticians of the Centuries* (Heyde and Seneta, 2001, p. 124). The same publication contains a wealth of other material about this remarkable man.

at the bus stop, we ask other passengers how long they have been waiting, their answers should give us some clue about how long we might have to wait. The longer the gap since the last bus, the less time we should expect to wait before the next one. If the buses are keeping to the timetable, information on recent arrivals should tell us a good deal. If, for example, we arrive as a departing bus recedes into the distance we can expect a long wait.

Maybe we can approach the idea of randomness in time by thinking about what we would need to know in order to make some prediction about when the next event will occur. Any information about when buses have arrived in the recent past should provide some useful data. The less the information, the closer we seem to be to randomness. Perhaps, then, we should define randomness in time as a state of affairs in which all such information would be useless. This is entirely consistent with my earlier definition of a pure-chance event as one where there is nothing we can observe which will help in its prediction. Put another way, there is no discernible reason for what happens.

Roughly speaking we can formalise this by saying that the expected waiting time (for the next bus, for example) is the same, however long since the last one passed. Whether we have been waiting for one minute or half an hour makes no difference to our prediction. More exactly, this can be expressed mathematically to tell us what the frequency distribution of inter-arrival times should be if this were true. It turns out to be what is called the *exponential* distribution. It looks like the frequency distribution sketched in figure 3.1.

This distribution has its highest frequency near time zero and declines steadily as the waiting time increases.[3] Short

---

[3] Students are often confused by what seems to be an implied contradiction about the distribution of waiting times. The frequency does decline as stated,

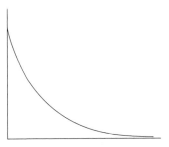

Fig. 3.1 The form of the exponential frequency distribution

waiting times are, therefore, much more common, relatively speaking, than large ones. This distribution occurs whenever the condition for randomness is met, so we might legitimately refer to it as a law. It is a regular and universal occurrence under specified conditions – and that is what laws usually are – but it is a law generated by randomness, which seems rather odd because we tend to think of laws as determining what will necessarily happen. Here we suppose that there are no determining factors, and yet we have a law.

But this is not the end of the story, because this exponential law is not actually what is usually known as Poisson's law. The latter arises if we describe the process in another way. Instead of observing the times between events we could count how many events occur in fixed intervals of time. Standing at the bus stop we could count how many buses arrive in successive ten minute intervals, say. The result would be a string of numbers which might be something like the following:

yet the expected waiting time is constant no matter how long one has waited. The first relates to someone newly arrived, the second to someone who has already waited some time. When you join a queue the chance of having to wait time t certainly decreases as t increases. But once you have been in the queue for a length of time t your position is no different from that of someone who has just joined.

3, 0, 1, 3, 2, 0, 0, 2, . . .

These do not look very lawful, but if we collect them together a pattern begins to emerge. In a long sequence the successive proportions will display a characteristic pattern known as the Poisson distribution or Poisson's law. In earlier times it was sometimes referred to as the Law of Small Numbers (in contrast to the Law of Large Numbers) because it showed that, even with rare events, there may be stable patterns.

I do not need to go into the mathematical details beyond saying that the proportion of 3s, say, is simply related to the proportion of 2s, and so on. The former is obtained by multiplying by the average and dividing by 3 (or whatever the current number is). There are simple tests which can be applied to see whether the proportions do conform to this pattern, and it is that pattern which constitutes the law.

In reality it is, as we have seen, just another way of describing randomness in time. There is no sense in which occurrences have to *conform* to the law. It is merely a description of what happens when there is no law.

At this point the reader may still be slightly misled by the bus example that I have used to illustrate these ideas. The objection may be raised that there are, in fact, reasons why buses arrive when they do. The fact that we are unable to predict them is because we lack the necessary knowledge. This is correct, and it illustrates the limitation of the illustration and also hints at serious theological issues. *Must* there be a reason, as in the bus example, or could it be there is no reason at all.

This brings us to the example of the emission of radioactive particles. These do seem to conform to the requirements of randomness in time with little prospect of our being able to look behind the phenomenon and discover what is going on. It is in such circumstances that some theologians have invoked

Fig. 3.2 The form of the normal frequency distribution

God as the controller. After all, if every event must have a cause, and if no other cause can be detected, the only alternative seems to be God.[4] This is a matter to which we must return but the point being made here is quite simple. Even when we specify a process which seems to be totally random, patterns emerge which display all the characteristics we associate with the idea of law. This must have profound implications for what we mean when we speak of the lawfulness of the world and what it means to attribute those laws to God.

THE NORMAL LAW

The oldest and best known of all statistical laws is the so-called Normal Law or, reflecting its origin, the Normal Law of Error. This goes back to Gauss but is now known to be so fundamental and widespread that its historical origins are incidental. The Normal Law[5] refers to the shape of a frequency distribution often described as bell-shaped. Figure 3.2 illustrates the shape.

[4] This would be the view of all those who take a broadly Calvinistic position.
[5] The Normal Distribution is often referred to as the Bell Curve, a term made popular by the somewhat notorious book with that title by Richard Herrnstein and Charles Murray (Free Press Paperbacks, 1994) on intelligence and class structure in American life. Ironically the name is somewhat tangential to the main message of that book.

A quantity which is normally distributed will have a frequency distribution like that shown above, where most values
are clustered in the middle of the range and then tail off towards
the extremes. It occurs widely in nature – human heights are
one example. Is this because it was built into the creation as
something to which a wide range of phenomena would be
required to conform? The answer is, almost certainly, no. It
may be thought of as another example of order out of chaos.
It simply describes one of the regularities of randomness. In
other words it takes a world of chance to produce this law.

A simple example will help to illustrate one way in which
the law comes about. As always in this book, I shall ignore the
technical conditions which must be imposed if the mathematics
operating behind the scenes is to deliver the results. For present
purposes this is of no practical significance and enables us to
penetrate to the essential idea with greater ease.

Imagine that we are interested in the age distribution of the
members of a human society. Because they are too numerous
to question individually, we select a random sample. (A fuller
discussion of random sampling is given in chapter 10.) A random sample is one chosen so that all possible samples have an
equal chance of being selected. In other words it is the fairest
possible method of selection – no sample has any advantage
over any other. If the sample is fairly small – under a hundred,
say – it will not be too difficult a job to ascertain all the ages.
Suppose we do this and find the average. Now imagine this
to be repeated many times on different, independent, samples
with the average being computed on each occasion. Surprisingly the frequency distribution will have a form very close to
the Normal Law. The remarkable thing is that we can say this
without any knowledge whatsoever of the population age distribution. The result must, therefore, be due to the method of
drawing the samples and have little to do with the underlying

distribution itself. It is thus a product of the sampling mechanism, and human choice had nothing to do with it nor did it depend on any other feature of the world. Order, in the shape of the distribution, has come out of the randomness of the sampling procedure. It tells us nothing about any Creator but more about our own ingenuity – or does it?

To take this example a little further, I return to the emission of particles from a radioactive source. If we observe many emissions we shall find that the frequency distribution of the time to the next emission is highly skewed. As already noted, short intervals will be the most common and the frequencies will decline as the duration increases. This is nothing like the normal distribution, but the normal is not far away. Suppose we repeatedly form the average of, say, twenty emission times and look at *their* distribution. It will turn out that this is quite close to normal. Once again, this has little to do with the waiting-time distribution but neither, on this occasion, can we attribute it to the randomness of sampling, for we did not sample the intervals but took them as they came. This time the randomness lies in the variation in waiting time and the *independence* between successive emissions.

The surprising thing here is that the form of the waiting-time distribution has very little to do with the result. It would have been much the same whatever that distribution had been. The unpredictability of the individual emission turns into the predictability of the final distribution, with minimal conditions on what the initial distribution is. It does not even matter whether all the distributions are the same nor is it vital that they be totally independent.

The ubiquity of the normal distribution arises not so much because it is built into the structure of the world but rather that it is imposed onto the world by the way we choose to observe it. It may, therefore, tell us very little about the intentions of the Creator. But it does tell us that order and randomness are so closely tied up together that we can hardly speak of one without the other. Total disorder seems impossible. Order is discernible at the aggregate level in even the most chaotic micro-level happenings.

DYNAMIC ORDER

Order can be more subtle than that considered so far. It can refer to stable *patterns* which develop and persist over time. Stuart Kauffman[6] has explored this kind of order over many years in an effort to discover how biological complexity might have emerged. I shall later dismiss the common simplistic arguments about the origin of life which imagine all the ingredients shaken up together and linking themselves together at random. This kind of process bears no relation to any kind of plausible biological process. Kauffman's approach may be crude but it is far more realistic in that it allows the interaction among the ingredients to play their part. Kauffman's aim is not to model biological processes in minute detail – that would be quite

[6] Stuart Kauffman is a theoretical biologist who has championed the view that the self-organisation of the relevant molecules may have created the complexity necessary for life to appear. This has attracted the opposition of Overman because it would mean that life would have appeared spontaneously, without the need for any Designer. Kauffman, himself, sees his own approach as meeting the objections of those who claim that life could not have arisen 'by chance'. For example he draws attention to the extremely small probabilities for the origin of life given by Hoyle and Wickramasinghe 1981 (see Kauffman 1993, p. 23) as showing that such a random assembly was impossible. Kauffman sees such calculations as irrelevant. If we reject special creation, this leaves the way clear for his own approach. There are other, chemical, arguments for believing that life might arise spontaneously in the conditions obtaining on the primeval earth. Recent examples are reported on page 12 of *New Scientist*, 18 January 2003 and page 16 of 16 December 2006.

impossible – but to capture the essential qualitative features of this system. This may depend on the broad characteristics of the system rather than its details. He constructs a model consisting of elements linked together in a random fashion and which exert an influence on one another via links. In reality the elements might be genes or enzymes but, at this stage, it is the broad qualitative properties of such systems that I wish to study. Many results for systems of this kind are given in Kauffman (1993) or, at a more popular level, in Kauffman (1995).

For illustrative purposes Kauffman imagines the elements as being electric light bulbs which can be on or off. At any one time, therefore, there will be a display of lights formed by those bulbs which happen to be on.

The pattern now changes in the following manner. To take a particular case, let us suppose that each bulb is connected to three other bulbs, selected at random. The state of the initial bulb at the next point in time is supposed to depend on those of the three bulbs to which it has been connected. There has to be a rule saying how this works. One possibility is that the initial bulb will only light if all those to which it has been connected are also alight. What has been done for one bulb is done for all, so that we end up with a network of connections, such that the state of any bulb at the next step is determined by the three to which it was connected. Such a system is set off by having an initial set of bulbs illuminated. Successive patterns will then be generated by the on/off rules. A network of this kind is called a Boolean[7] net because the two sorts of rules correspond to the *and* or *or* rules of Boolean logic. One or

---

[7] George Boole (1815–64) was a British mathematician who is remembered as the originator of Boolean algebra, which is central to computers which work with binary numbers.

other of these rules is attached to each bulb for determining whether it is switched on or off.

The question now arises: what will happen if such a system is allowed to run indefinitely? The first thing is to note that there will usually be a very large number of possible patterns. For example if there are 100 bulbs there would be 2^{100} possible patterns, which is an extremely large number. More realistic examples with, say, 1,000 bulbs, would be capable of generating an unimaginably large number of possible states. To run through the full gamut, even at the rate of one million per second, would take longer than the time the universe has been in existence! This might lead one to expect a seemingly endless succession of meaningless patterns.

The surprising and remarkable thing is that this may not happen. In some circumstances the system may reach a relatively small set of states around which it cycles for ever. This small set may have only a few hundred patterns which is negligibly small compared with the original set. It turns out that the critical numbers in determining the behaviour of the system are the number of bulbs and the number of other bulbs to which each is linked. We used three as an illustration. It turns out that if there is only one connection, the behaviour is rather uninteresting, whereas if it is large, greater than four or five, say, the behaviour is chaotic and there are unlikely to be any simple patterns. It seems that two connections are capable of producing more interesting behaviour, sandwiched between the monotony of just one connection and the chaos of many connections. It is on this borderland of chaos and order that the interesting things happen. There is just sufficient order to produce simple patterns which persist, but not so much as to render the result chaotic. Order on the edge of chaos seems to be a key characteristic necessary for the emergence of interesting phenomena.

The patterns here are not static. The order lies in the repeatability of a small set of states through which the system continually cycles. The order, remember, arises from a random specification of the connections. It does not require any detailed design work to produce the patterns. These persist in spite of the randomness in the structure. It can also be shown that other minor variations of the rules governing the behaviour leave the broad characteristics unchanged.

IT'S A SMALL WORLD

There are about six billion people in the world so it would seem obvious that almost all of them are so remote from us that we could have nothing in common. Yet there are so many surprising occurrences – as when we meet two people in different spheres who turn out to know one another, and we comment 'it's a small world'. This turns out to be truer than we might think when we start to wonder how we might measure our closeness to other people. There is a considerable literature on this and related phenomena to which Barabási (2003) and Buchanan (2002) provide a useful point of entry.

There are some well-known examples of the phenomenon. Two widely discussed cases relate to the media and mathematics. Kevin Bacon acted in films. Other actors who appeared with him in a particular film were obviously close to him in a certain sense. People who did not appear in any films with him, but who appeared with someone who *had* appeared with Kevin Bacon, are less close. Let us say that they are distance 2 from Bacon. Those who came no closer than acting with those with number 2, may be said to have distance 3, and so on. What is the distribution of these distances, or, what is a typical distance? It seems that the typical distance is around

6 which, given the number of actors and films seems to be a remarkably small number.

Mathematicians and many others whose work involves mathematics have what is known as an Erdös number. Paul Erdös was a prolific mathematician who collaborated with many other mathematicians. One's closeness to Erdös can be measured by one's Erdös number.[8] Those who co-authored a paper with Erdös have an Erdös number 1; there are about 509 of them. Those who collaborated with a collaborator, but not with Erdös himself, have an Erdös number 2; there are currently about 6984 of these and their names have been published. Almost every mathematician has a small Erdös number – usually in single figures. Again, this seems rather remarkable given the vast size of the published literature in mathematics.

It turns out that this phenomenon is widespread and in totally different fields, ranging from power grids to the study of infectious diseases. Is there something going on that we have missed or is this another example of order arising out of chaos? The chaos in this case is the haphazard collection of links which are formed between pairs of individuals. 'Haphazard' seems to be the right word to use here because it conveys a sense of lack of purpose. There may, indeed, be purpose in the formation of individual links, but it is the overall pattern with which we are concerned. It does not need anyone to oversee the overall process. Purely local connections of a haphazard nature are sufficient for the 'small-world' structure to emerge.

The study of small-world phenomena appears to have begun with an experiment conducted by Stanley Milgram[9]

[8] The Erdös Number Project has a website: http://www.oakland.edu/enp, from which this and similar information can be obtained.

[9] Stanley Milgram (1933–84) was a psychologist who served at Yale University, Harvard University and City University of New York. While at Yale

in 1967. His work was concerned with how far any of us might be from some public personage. We might ask whether we can find links through personal acquaintances which stretch from us to the said personage. Although many of the details of the experiment might be criticised, it established the hypothesis that the average number of such links was about six. If one wanted to establish contact, for example, with the President of the United States by going through 'someone, who knew someone, who knew someone . . .' it should not therefore need too many steps to do so.

ORDER IN NETWORKS

The small-world phenomenon is one example of what happens in networks created by linking individuals in a haphazard way. Networks[10] are everywhere and have become such a familiar feature of our world that the word has passed into the language

he conducted experiments on obedience to authority which attracted a good deal of criticism on the grounds that they were unethical. Also at Yale, in 1967, he conducted his 'small-world experiment'. This was an empirical investigation in which he sent letters to sixty randomly chosen people in Omaha and Nebraska. Recipients were asked to hand the letters to some one they thought might be able to reach the target – a stockbroker in Sharon, Massachusetts. The results were somewhat ambiguous and were subject to a good deal of criticism but it was a truly pioneering study and has proved a point of reference for much work since on the small-world phenomenon.

[10] There has been a great deal of interest in *networks* in the last few years. This has resulted from the discovery that many of the networks of real life share a common architecture. Much of this work has been done by physicists who have seen parallels between the physical systems with which they work and networks arising in the social sciences. Since about the year 2000 there have been several books published which aim to make the results obtained available to the wider public. Some, such as Strogatz (2003) and Barabási (2003), are authored by active researchers in the field. Others, such as Buchanan (2002), are by experienced popularisers who are able to take a

as a verb. In spite of their diverse origins, networks display many common features remarkable for their simplicity. They are a striking example of order out of disorder. Even highly random networks exhibit order.

The simplest example is appropriately called the *random net*. This appears to have had its origin in the study of neural nets, where the interest was in such things as paths through the network along which communication could take place. A clutch of papers was published in and around the 1950s in the *Journal of Mathematical Biophysics* by Anatol Rapoport[11] and others. It was quickly recognised that the results also had applications in the theory of epidemics, where the nodes are people and the links are paths of infection. This work was summarised in chapter 10 of Bartholomew (1982) (though rather more detail is given in the first two editions of that book, 1967 and 1973).

Rather surprisingly this work was not noticed by the mathematical community until a new impetus was given to it when Erdös and Rényi weighed in with a series of eight papers on random graph theory beginning in 1959. (The details will be found at the top of page 245 in Barabási (2003).) This, together with the growing power and accessibility of computers, opened the way for the rapid contemporary development of the field. An interesting example of the application of a random net is given by Kauffman in his book *Investigations* (2000, pp. 35 ff.). Kauffman's interest is in the chemistry behind

broad but informed view. On 13 April 2002 the *New Scientist* published an article by David Cohen entitled 'All the world's a net'.

[11] Anatol Rapoport (1911–2007) was born in Russia but emigrated to the United States in 1922. He has worked on a wide range of topics but the focus was on the mathematics of decision making. From 1970 he was based at the University of Toronto where he held chairs in Psychology and Mathematics and in Peace and Conflict Studies. It is surprising that his work on nets was unknown to Erdös and his associates.

the origin of life but what he calls 'the magic' comes across without going into the background.

Kauffman supposes that we have a large number of buttons scattered across a hard wooden floor. We begin by picking two buttons at random and joining them with a coloured thread. Next we repeat the process, joining another pair of buttons. It could happen that one or both of those picked the second time could include one of the original pair. This process is repeated again and again. Every so often we pause and pick up a single button chosen at random. It may turn out to be an isolated button or it may be linked to others which will also be lifted because of the threads which tie them together. In the early stages these linked clusters will usually be small but as time goes on they will get larger. One way of charting the progress of this clustering is to see how the size of the largest connected cluster increases as the number of threads (connections) increases (expressed as a proportion of the number of buttons). The magic occurs when this proportion comes close to a half. Below a half the size of the largest cluster will be small, but as it passes through a half it increases rapidly and soon comes close to the maximum possible. In chemical terms this is a *phase change*. We move from a situation in which groups are small to one in which connectivity is high. This happens without any 'design'. The connections are purely random and yet this haphazard sort of linking produces a situation which, to some eyes, might appear to express purpose or design.

It is worth reflecting on this example a little longer to see what these random connections have done for us. Suppose, instead, that we initially lay the buttons in a straight line. If we now connect neighbouring pairs we shall eventually create a single cluster and the number of links will be one fewer than the number of buttons. Expressed as a proportion, the ratio of links to buttons is very close to one – much higher than was obtained using random connections. The clusters obtained by the two methods are likely to be very different and for some purposes one or the other might be preferred. The point is that it does not take many links, randomly assigned, to create something very close to a single structure.

Networks can be created in all sorts of ways and the features of interest may vary greatly. What does not vary is the presence of patterns or regularities which seem to appear in spite of the chaotic way in which things happen.

Epidemics provide another example. Infections such as influenza spread in human populations largely through contacts. Who meets whom may be fairly random as in business, shopping, education, and so on. Contact, in individual cases, may be planned or unplanned yet the spread may be very rapid. Diseases such as AIDS spread through sexual contacts. Rumours spread in a manner rather similar to infections. They are passed from friend to friend and often, it seems, at lightning speed. A sufficiently hot piece of news will spread quickly enough without anyone going to the trouble of engineering the spread. This fact is well known by those who wish to counter rumours. To counter such a rapid spread, broadcasting in some form which reaches everyone almost immediately is essential.

There is a very extensive scientific literature on how epidemics, whether of infection or rumour, spread. Not surprisingly, some things about such processes are predictable and constant in spite of the fact that they are driven by purposeless activity. In the simplest examples, anything spread in a given population will eventually reach everyone if enough time is allowed. This is because in a freely mixing population everyone will meet everyone else at least once and possibly several times. But not many epidemics are like this; usually there is some self-limiting factor. People who become infected, for example, will withdraw from the population to recover. This is voluntary isolation but, in serious cases, such as SARS,[12] they will be removed from the exposed population. Rumours, similarly, often die out before the epidemic has run its course. Passing on a piece of salacious gossip ceases to be attractive if potential hearers have already heard. This may cause the spreader to give up.

Models have been constructed to quantify some of these intuitively based observations. Under suitable conditions it is possible to predict the proportion of those who have heard or been infected at any stage of the epidemic. The growth curve also has a characteristic shape, rising slowly at first, then more rapidly before finally levelling off. This shape also depends remarkably little on the precise details of how the spreading takes place. What is interesting for present purposes is that haphazard and seemingly unguided processes prove to have more pattern than we might have guessed. Randomness achieves easily that which, by design, might have been very difficult.

In all of this work it is supposed that the population of nodes (e.g. people) is fixed. In practice many networks are constantly growing; the World Wide Web is, perhaps, the prime example. It is found that such networks differ in some important respects from the random net. For example, in a random net the number of incoming paths to a node will vary but not by much. The actual numbers will have a distribution fairly tightly packed around the average. Growing networks are not like this. They have a number of what are called *hubs* which are at the centre of large numbers of links. To account for this Barabási and his colleagues proposed a model leading to what they called

[12] Severe Acute Respiratory Syndrome.

scale-free networks. They supposed that as each new node appeared it established links with already existing nodes. The choice of destination could be purely at random or the probability could be proportional to the number of links already established. It is intuitively clear that nodes which appeared near the beginning of the process will tend to have more links, simply because they have been around longer and so have had more opportunity. The effect of growing in this way is to create a hierarchy of hubs, some with very large numbers of links. The distribution of the number of links per node now becomes highly skewed, with large numbers having very few links and a few with a very large number. This pattern is preserved as the system grows and hence accounts for the description *scale-free*. For our purposes it is important to note that a large measure of randomness is still present in the formation of the net but the resulting pattern, though still simple, is very different from that of a random net. It is also interesting to note, in the context of this book, that one of the examples Barabási uses is the spread of early Christianity through the activities of St Paul, who created a network as he travelled between centres of population in the ancient world.

The number of links per node emerges from these analyses as a key descriptor of a network. Once we get away from the random net, the distribution of that number is, as we have seen, highly skewed. But more can be said about it than that. The long upper tail of that distribution can be approximated by what is known as a power law. Such distributions have been known for a long time in other contexts of which the best known is, perhaps, the Pareto Law of Income Distributions. They are yet another example of order out of disorder. Much of that work is brought together in chapter 7 of Bartholomew (1982).

APPROXIMATE ORDER

Appearances can be deceptive. Order can appear, persist for a time and then disappear. I describe a simple example merely to illustrate the possibilities. In a sense, the example is like a sequence of independent tosses of a coin with the difference that the probability of a head at any toss changes from time to time depending on the outcomes of previous tosses. The experiment goes as follows. Toss a fair coin 100 times, say, and compute the proportion of heads. Make a further 100 tosses but this time make the probability of a head be equal to the proportion of heads in the previous 100 tosses. Repeat this procedure and at each stage let the probability of a head be equal to the proportion of heads in the previous set of 100 tosses. What will happen in the long run to the probability? Intuition might lead us to expect it to remain around 0.5. We expect to get about 50 per cent of heads the first time round that should lead to around the same figure next time, and so on. In the very long term this argument will let us down. Sooner or later all of the 100 tosses will be all heads or all tails. When that happens the future is fixed, because if we estimate the probability to be 0 or 1 we shall get nothing but all heads or all tails for ever. Nevertheless, in the medium term, the probability is likely to remain around 0.5 for some time, thus exhibiting a degree of order which, though giving the appearance of constancy, is doomed to disappear. Thus order, when it arises, is not necessarily something which is permanent.

SYNC

Christiaan Huygens invented the pendulum clock in about 1655. Ten years later he suffered a slight indisposition and was in a room with two such clocks when he noticed 'a marvellous thing' about which he wrote to his friend, Sir Robert Moray.[13] He noticed that two similar clocks, hanging together in the same room, were beating in time. This was an early and very simple example of synchronisation. This might have happened because the clocks were set going together and, being very accurate timekeepers, had kept in step ever since. But Huygens found that, however they were started, they would always synchronise in about half an hour. There must have been some means by which the action of one clock influenced the other in a manner which brought them into harmony.

Pendulum clocks are not the only things which interact in this way. Synchronisation is a very widespread phenomenon and includes some spectacular examples. Fireflies, as their name implies, produce flashes of light and these occur at regular intervals. If one encountered a large number of flies, all flashing, one would expect to see a chaotic mixing of the individual flashes but this is not what happens. In Southeast Asia and elsewhere thousands of fireflies gather along river banks and provide a spectacular display of synchronised flashing. Sometimes this type of phenomenon can be harmful, as in epilepsy, when the synchronised activity of many brain cells can lead to convulsions. At other times it can be essential as when many pacemaker cells in the heart combine to produce a regular heartbeat. This kind of phenomenon is remarkably widespread, which suggests that there is a common pattern in what is going on which might be susceptible to mathematical analysis. This turns out to be the case and the topic is the subject of a fascinating account by Steven Strogatz, one of

---

[13] Further details of this example and some further background will be found in Strogatz (2003, chapter 9).

the principal researchers in this field. (See Strogatz (2003) for example.)

This phenomenon is certainly an example of spontaneous order arising from disorder. It is not necessarily a case of order arising from chance, so some justification is required for its inclusion here. The essential point is that there is no overarching control which produces the order represented by the regularly beating heart. There is no central timekeeper or conductor orchestrating the behaviour of fireflies. The pattern arises from local interactions which bear no obvious relationship to the global pattern. The clocks will beat in time however much or little their starting positions differ. It would make no difference if the starting positions were determined randomly and that means that this aspect does not have to be designed or planned. Viewed theologically, the regularities can be achieved without any detailed planning and hence there needs to be no detailed control at the micro level. Spontaneous synchronisation can be achieved almost for nothing.

CONCLUDING REMARKS

I have ranged widely over many fields to show that order often arises as a consequence of disorder. The 'laws' I have considered might speak to the uninitiated of the direct intervention of a God, at every stage, to ensure that things turned out exactly as he desired. In fact the beautiful shape of something like the Normal Distribution might be regarded as eloquent testimony to the nature of the Designer. But as Mark Twain memorably remarked, 'it ain't necessarily so'. Lawful behaviour in the shape of statistical laws seems to emerge naturally from an underlying chaos. Indeed one might go so far as to say that chaos is a precondition of the order which, to some, speaks so eloquently of the divine mind.

The full theological implications of all this are profound but a discussion of them must wait until more pieces of the jigsaw are in place. For the present, I note that the involvement of God in the creation may be much more subtle than theologians, rushing to premature closure of the science–religion debate, are prepared to allow.

CHAPTER 4

Chaos out of order

The transition from chaos to order is not a one-way process. Just as order can result from chaos, so can chaos result from order. In this chapter I describe three ways in which this may happen. These are: accidents, pseudo-random numbers and mathematical chaos. This transition, also, has to be taken into account by our theology. If these were the only sources of uncertainty in the world there would be no need to invoke pure chance or to explain how it might be consistent with divine purpose.

DISORDER GENERATED BY ORDER

Our theological assertions about what God can or cannot do in the world depend very much on what kind of place we believe the world to be. In the last chapter we encountered the somewhat disconcerting fact that much of the order and lawfulness which we so readily attribute directly to God has its roots in disorder. But this is only half of the story – a great deal of disorder attends the regularities that are all around us. In this chapter, therefore, we shall look at the other side of the coin as a prelude to seeing whether the apparently paradox-ical situation which faces us can be resolved. For example, if the purposefulness of God is to be discerned anywhere, one would expect it to be in the regularity and predictability of the aggregate. But, if this is the case, how does God engineer it?

On the other hand, if the motions of individual molecules in a gas are to be thought of as purposeless and undirected, how can we attribute purpose to the aggregate behaviour which is built on this irregularity?

This will demand a reassessment of what it means to say that happenings in the world reveal the purposeful activity of God – or the lack of it. Here we shall look at three manifestations of chaos out of order: first, accidents, which in retrospect can often be seen as determined by antecedent events but which constantly surprise us; secondly, pseudo-random numbers – this may seem to be an esoteric branch of mathematics but, in reality, it is very big business in applied science of both the natural and social varieties; finally, chaos theory, which, in a more technical sense, has come into great prominence in recent years and has sometimes been described as the 'third great revolution in physics this [the twentieth] century'. Some theologians have seen chaos theory as a lifeline offering a means for God to exercise control without disturbing the orderliness of nature. To me this claim seems premature and unconvincing.

ACCIDENTS AND COINCIDENCES

Accidents often appear to be unintended and unpredictable happenings. The phrase 'accidents will happen' expresses both their inevitability in the long run and the uncertainty surrounding particular occurrences – as does the remark about 'accidents waiting to happen'. Speaking of 'happy accidents' reminds us that it is not the consequences of the happening that define an accident. The essential idea behind the word is of something that is not planned or intended. There are no obvious causative factors and, though there may well be reasons for accidents, they are largely hidden from us. In other words, they bear all the signs of being chance events. Yet they can happen in systems that are fully determined.

Coincidences are closely related to accidents. Again there is no suggestion that they are unnatural in any sense; two events just happen to have occurred at the same time and place. The distinctive thing about them is the juxtaposition of two unrelated events which take on a new meaning because of their coincidence.

The striking of the earth by a large asteroid may, except perhaps in the last stages of its approach, have all the appearances of an accident in the sense that it was unintended and largely unpredictable. But we know that heavenly bodies are subject to known gravitational forces and, given their trajectories over a period, collisions could be predicted with certainty. The whole set-up may well be deterministic but our ignorance of the contributing paths of causation would lead to our being just as surprised as we would have been by some truly random happening. From the perspective of chapter 2 on 'What is chance?' we simply do not have enough information to predict with certainty what is going to happen; we are therefore uncertain.

At a more mundane level, accidents also commonly arise from the coincidence of two independent paths in space and time. In speaking of paths we are borrowing terminology which is familiar enough in two-dimensional geometry but may seem odd in this context. Just as we can plot a path on a two-dimensional map, so we can imagine doing it in more dimensions than we can actually visualise. With three space dimensions and one time dimension we have a four-dimensional space but we can stretch our language to speak of a path in the same way as in fewer dimensions where we can visualise what is going on. When two causal paths coincide at a particular time and place they may trigger the event we call an accident. To an all-seeing eye the accident could be foreseen and would come as no surprise. We sometimes find ourselves in this position. We may shout a warning to someone who cannot see what is coming. To them it comes as a surprise because they do not have enough information to make the prediction that is a certainty to us.

An accident could, of course, be a truly random event triggered by some purely chance happening as, for example, the emission of a radioactive particle. Usually, however, it will only appear to be random because of our ignorance of the processes which give rise to it. All that we need to establish at the moment is that, even in a fully deterministic system, events happen that have, to us, all the characteristics of chance events. For all practical purposes accidents and coincidences are just like pure-chance happenings, but the theological issues they raise are quite different.

At one level accidents pose no serious theological questions. If from God's perspective all is known to him in advance, then his sovereignty is unchallenged. At another level, of course, accidents raise profound questions as to why God should allow them to happen. In turn, this leads on to the question of whether the system could have been designed to avoid all such undesirable happenings. Given the highly complex and interacting nature of the creation, it is not at all obvious that it would be possible to 'design out' all unwanted outcomes. The highly interconnected character of the world may mean that one cannot have some desired outcome without having others which, so to speak, necessarily go along with it. This is a problem for those who demand total sovereignty for God at every level. It may be less of a problem for those who are prepared to allow the elasticity which a modicum of uncertainty provides.

PSEUDO-RANDOM NUMBERS

Pseudo-random numbers are big business. The production of such numbers by computers – even pocket calculators – is a largely unnoticed yet universal feature of contemporary society. Basic scientific research and the selection of winners in competitions both call for a ready supply of random numbers. But what are pseudo-random numbers, why do we need them, and what bearing does their generation have on the present question of chaos out of order? In passing, it should be noticed that genuinely random numbers can be produced using physical processes, such as radioactive decay. If this is so and if the real thing is available one might ask why we should bother with something which is second best. The answer is that pseudo-random numbers are almost as good and do not require any physical apparatus with all its costs and inconvenience.

It may help to begin by looking at some familiar things which serve to introduce the general idea. In some sense we are talking about what might be called 'contrived' accidents. Think first of the tables of figures published by National Statistical Offices. Narrowing it down, imagine we are looking at the populations of large cities in a big country such as the United States. One such figure might be 1,374,216. Not all of the digits have equal importance. The first 1 is the most important. It immediately gives us an idea of whether this is a big city or not; in this instance it is in the million or more bracket. The next most significant digit is the 3 in the second position. Knowing that the city population was in the region of 1.3 million would probably tell us most of what we want to know and it would certainly fix the size in relation to most other major cities. As we move along the sequence the digits become progressively less significant. Another way of putting the matter is to say that the information, conveyed by these successive digits, diminishes as we move along the sequence. The final 6 tells us very little indeed about the size of this city, or about its size in relation to other cities. In fact, it may not be accurate given the uncertainties of counting such things. If we were to collect together the last digits of the populations of many cities they would be almost devoid of meaning and so would be rather like random numbers. But we very well know that they are not strictly random. They have been arrived at by a process of counting well-defined objects (human beings, no less). It would not be too far-fetched to describe collections of digits arrived at in this way as pseudo-random digits on the grounds that while they are certainly not random, nevertheless they look very much as if they were.

Any method which involves using some well-defined arithmetical process and which then discards the most meaningful parts is a way of producing pseudo-randomness. One traditional method, long used by children, is the use of counting-out rhymes to decide who shall be *it* in a game. A method is required which is accepted as fair by the participants in the sense that, over the long run, it does not appear to favour any particular individual.

There are several variations of the method but the usual procedure is to count around the circle of children, counting one word to each child. The child on whom the last word falls is selected. In principle it would be perfectly possible to work out in advance who would be selected. To do this we would divide the number of words in the rhyme by the number of children in the group. We would then disregard the integer part of the answer and keep the remainder. This number determines which child would be selected. If the remainder was four it would be easy to spot the fourth member of the circle – much easier than doing the full mental calculation. The method works because it is not immediately obvious what the remainder will be, because people do not carry in their heads the numbers of words in all, or any, of the many rhymes there are. It is much easier to remember the words than their number. The remainder is a pseudo-randomly selected digit.

For the large-scale uses of random numbers these simple methods would be far too slow, their yields would be inadequate and there would be serious doubts about whether they were sufficiently random. These difficulties are overcome by using what are called random-number generators. Many of these are based on much the same principle as the counting-out rhyme. One does a calculation, such as a division sum, and then retains the least significant part of the answer. Nowadays most computers incorporate random-number generators. One characteristic of many such generators is that, if you go on long enough, the output will start repeating itself. This is one sign that the numbers are not genuinely random but this fact is of little practical importance if the number we need is very small in relation to the cycle length.

It is worth observing at this stage that coin tossing is also pseudo-random. It does not produce a sequence of digits from 0 to 9 but it does produce the simplest kind of numbers we can have – that is binary numbers. If we code a head as 1 and a tail as 0, a sequence of tosses will be a string of 0s and 1s. Whether or not a coin falls heads or tails depends on what is left over when the number of complete rotations is finished. This, in turn, depends on the impulse and spin imparted to the coin by the tosser's hand. Whether or not it falls heads or tails depends on which side of the vertical the coin is when it hits the ground on its return. The binary outcome is thus determined by a remainder to the number of tosses in much the same way as are pseudo-random numbers. This comparison shows just how close to being genuinely random a pseudo-random sequence can be. In turn, this raises the question of whether there are degrees of randomness and whether they would tell us anything of theological relevance.

There are two entirely distinct ways of approaching this question. The traditional statistical approach is to ask what properties a random series should have and then to examine a candidate series to see whether it has them. In this approach there is no such thing as 'a random number', in the singular. The randomness refers to the *process* of generation and can only be detected in a long sequence of such numbers. A pseudo-random-number generator is a device which produces a sequence of numbers (usually the digits 0, 1, 2, . . . , 9 or the binary digits 0 and 1), which for all practical purposes are indistinguishable from a purely random sequence. And what would a purely random sequence look like? It would look like a sequence from a purely random generator! We might begin to specify some of the required properties. To begin with, the number occurring at any point in the sequence should not depend on anything that has gone before. If it did, it would be, in part, predictable. More generally, nothing we could learn from part of the series should contain any information for predicting any other part. This all depends on asking about the mechanism which generated the series. According to this approach one does not ask whether a particular string is random. *Any* string whatsoever *could* have been generated by a random process. A sequence of 20 sixes generated by rolling a six-sided die has probability $(1/6)^{20}$ which, although an exceedingly small number, is exactly the same probability as any other sequence, however random it might look. Instead one asks whether the generating mechanism was such that all possible digits had the same probability of occurring independently of all other outcomes. From this one can deduce various properties which a series should have. For example, in the binary case, each digit should appear roughly equally often; each time a 1 occurs it should be followed equally often by a 1 or 0, and so on. By making these comparisons one can never categorically rule out any series as non-random but one can get closer to saying whether it is likely to have been generated by a random generator.

The second approach is to ask how close the actual series is to randomness. This is based on a clever idea of Gregory Chaitin[1] (see his chapter 2 in Gregerson, 2003a) though, like so many other good ideas, it had been anticipated some years earlier. To avoid confusion it would be more accurate to describe the method as measuring the amount of *pattern* rather than the degree of randomness. A highly non-random series exhibits a large degree of pattern and conversely. For example a sequence of 100 ones is highly patterned and it can be described very concisely as 100 ones. The sequence 01010101 . . . is almost equally simple and could be described as 01 repeated 50 times, say. Chaitin's idea was that the simpler the pattern, the fewer the words which were needed to describe it. At the other extreme, randomness occurs when there is no detectable pattern at all and then there is no shorter way of describing it than to write the sequence out in full. Actually one has to be a little more precise than this. Descriptions have to be in terms of the length of the computer program (on an ideal computer) necessary to reproduce the sequence. Nevertheless the idea is essentially very simple. Randomness is measured by the degree to which one can describe the series more economically than by using the series itself. We shall meet this second way of measuring randomness again in chapter 7.

CHAOS

In ordinary usage the word chaos carries a much wider connotation than described in this section and this tends to spill over into theological discussions. I am talking here about what

---

[1] Gregory J. Chaitin, born in 1947, is a mathematician and computer scientist who apparently had the basic idea mentioned here while still at school. There are many accounts of this way of measuring randomness but the chapter referred to here (entitled 'Randomness and mathematical proof') has the merit of being written by Chaitin himself.

is more accurately described as *mathematical chaos*.[2] In fact it is, at least, debatable whether anything exists in the real world corresponding exactly to *mathematical* chaos.

Science has been particularly successful where it can predict change over a long period. Tide tables are published well in advance because tides depend on accurate knowledge of the relative positions of the earth, moon and sun and how they change with respect to one another. Similarly the great successes of engineering are the result of good knowledge of materials and how they respond to the strains and stresses that are put upon them. The study of dynamical systems has also depended on being able to determine how the rates of change depend on the current state of the system.

Mathematical chaos had its origins in the discovery that there are physical processes which do not appear to behave in a regular and predictable way. This is especially true of the weather. Forecasters attempt to model weather systems. They measure such things as air and sea temperatures at many points on the earth's surface and use models to predict how they will change and then construct their forecasts. Yet despite the growing complexity of their models and vastly increased computing power available it has proved impossible to look more than a few days ahead. Edward Lorenz[3] was one of the first to study the situation mathematically and to discover chaotic behaviour in an apparently deterministic model. He wrote down a set of differential equations which, he thought,

[2] There are many books on chaos aimed at different audiences. A popular account is given in *Chaos: Making a New Science* (1987) by James Gleick. An excellent, somewhat more technical account – but still not making undue mathematical demands – is *Explaining Chaos* (1998) by Peter Smith.

[3] Edward Lorenz (1917–) was a meteorologist working at the Massachusetts Institute of Technology. It was in simulating weather patterns that he discovered, accidentally apparently, that the development of a process could depend heavily on the initial conditions.

provided a reasonable approximation to real weather systems. As he projected them into the future it became apparent that the 'futures' which they generated depended critically on the precise starting assumptions. Thus was born chaos theory.

A characteristic feature of chaotic processes is this extreme sensitivity to initial conditions. It is this fact which makes long-term forecasting almost impossible and which justifies the introduction of the description *chaotic*. Two systems, which are extremely close together at one time, can subsequently pursue very different trajectories. This near lack of predictability is the source of the term chaotic.

Weather is not the only phenomenon which exhibits this sort of behaviour. Heart beats and the turbulent flow of liquids are two other examples which have attracted attention. It is not, perhaps, immediately obvious whether chaotic processes such as I have described bear any relationship to the pseudo-random-number generators discussed above. They do in the sense that one can choose simple indicators charting the path which a chaotic process takes which are, essentially, sequences of digits. In this form, their properties can be compared with sequences of pseudo-random numbers.

This similarity to random processes has certain attractions for theologians. For example, for those who are uneasy with any true randomness and hanker after the certainties of determinism, it holds out the prospect of a fully controlled universe with a fully sovereign God. For if the unpredictability of some dynamic processes displays the lack of predictability of chaos then, maybe, all the uncertainties of life are rooted in mathematical chaos. This, as I note elsewhere (for example in chapter 12), poses problems of another kind for theology, but at least it avoids the most flagrant introduction of pure randomness.

A second attraction of chaos theory for theologians is that it offers the prospect of space for God's action in the world which does not violate the lawfulness which is so integral to science. For if such different paths are followed by processes starting from positions which differ infinitesimally, then perhaps God can engineer changes of direction at the human level by such minuscule adjustments that they are not humanly detectable. This view of things creates enough space for God to act without the possibility of detection by humans.[4] Like so many seemingly simple recipes, this one raises enormous problems which we shall come to later. Note here, however, that it makes the major assumption that the world can, in fact, be controlled in this way and secondly, that it implies things about the nature of God from which the orthodox might shrink if they were made fully explicit.

For the moment I simply add chaos theory to the list of deterministic processes which contribute to the unpredictability of the world.

---

[4] The idea that chaos theory provides space for God to act without violating the lawfulness of nature is particularly associated with the name of John Polkinghorne. He clarifies his views in his contribution to Russell *et al.* (1995) (pp. 147–56, see especially the section on Chaos theory, p. 153). In the same volume, p. 348, Nancey Murphy makes a similar point but it should be emphasised that the views of both authors are more subtle than a superficial reading might suggest. We return to the subject of God's action in the world in chapters 8 and 9.

CHAPTER 5

What is probability?

Theologically important arguments sometimes depend on the correctness of probability calculations. In order to evaluate these arguments it is necessary to know something about elementary probability theory, and in particular, about mistakes which are often made. In this brief, non-technical, chapter I introduce some basic ideas of measuring probability and then identify several common fallacies. These involve the importance of the assumption of independence, conditionality and the so-called prosecutor's fallacy.

TWO KINDS OF PROBABILITY

Chance is the general-purpose word we use when we can see no causal explanation for what we observe. Hitherto I have not needed to quantify our uncertainties but we shall shortly move into territory where everything depends on exactly how large these uncertainties are. I have already spoken of probability in unspecific terms but now I must take the next step. Probability is a numerical measure of uncertainty and numerical probabilities have been used extensively in theological discussions. As we shall see shortly, there are events, such as the appearance of life on this planet, for which the way in which probabilities are calculated is absolutely crucial. Certain events are claimed to have extremely small probabilities and on their validity hang important consequences for what we believe about life on earth or, even, the existence of God.

Many of the arguments deployed are often little more than rather wild subjective opinions, so it is vital to be as precise as possible about probability statements. There is no need to go into technicalities but it is important to be aware of some of the major pitfalls and to be able to recognise them. The purpose of this short chapter is to set out the more important things that the reader needs to know before we come to such contentious issues as Intelligent Design, where the correct calculation of the probabilities is crucial.

Probability theory might seem to be one of the more esoteric corners of the field which can be passed over without losing the general drift of the argument. But much can turn on whether probabilities are correctly calculated and interpreted. This is just as true in fields outside theology, so I begin with an example of considerable public interest from forensic science to illustrate this point. The outcomes of trials and subsequent imprisonment have depended on probabilities. Sally Clark, a solicitor and mother of two young children, was imprisoned for the murder of her two children in 1999.[1] The prosecution argued that the probability of two cot deaths in one family was so small that it could be ignored in favour of the alternative explanation of murder. The probability calculation played a major part in the conviction of Sally Clark and her subsequent release after two appeals. DNA matching is another field where

---

[1] This case raised a number of important issues about the presentation of scientific evidence in courts of law. In this particular case the expert witness was a distinguished paediatrician and his evidence was crucial in the court's decision to convict Sally Clark of murder. It was subsequently recognised that he was not competent to give expert testimony on statistical matters. The Royal Statistical Society issued a press release pointing out that a serious mistake had been made and urging that steps should be taken to use appropriate expertise in future. In the United Kingdom the Crown Prosecution Service has issued guidelines for expert witnesses in such matters as DNA profiling.

it is becoming increasingly common for probabilities to be used in criminal cases. Probabilities of one in several millions are not uncommonly quoted in court as supporting evidence.

A probability is a number purporting to measure uncertainty. It is commonly expressed as a number in the range 0–1, with the upper end of 1 denoting certainty and the lower end of 0 corresponding to impossibility. A probability of 0.5, which is halfway, means 'as likely as not'. Anything larger than 0.5 is then more likely than not. The betting fraternity do not often use probabilities in this form but use odds instead. These are an equivalent way of expressing uncertainty and are, perhaps, more widely understood. Odds of 3 to 1 in favour of some outcome is saying the same thing as a probability of 0.75 and 'evens' is equivalent to a probability of 0.5. Odds tend to be favoured when probabilities are very large or small. In Sally Clark's trial the estimated odds of a cot death were 1 in 8543, a figure whose import is, perhaps, more readily grasped than the corresponding probability of 0.00012. Odds are, however, less convenient when we come to make probability calculations or to be explicit about the assumptions on which the probability is calculated.

But where do the numbers in which such probabilities are expressed come from? For our purposes we can identify two sorts of probability.[2] First there are those based on frequencies. These are used by actuaries when constructing life tables for

---

[2] This twofold division of probability is, of course, somewhat oversimplified. Elementary textbooks often begin with 'equally likely cases' and do not get much beyond problems concerning coins, dice and cards, where the possible outcomes are, self-evidently, equally likely. Traces of this idea appear in what we shall meet later where complicated biological entities are mistakenly treated as what we shall call *combinatorial objects* (see chapter 7, p. 110). What are here called degrees of belief can be further subdivided into objective and subjective (or personal) probabilities.

insurance purposes. If, for example, 60 per cent of forty-year-old men survive to their seventieth birthday we may say that the probability that a randomly selected forty-year-old man reaching that age is 0.6. Armed with such probabilities one can go on to estimate the probabilities of certain other events concerning forty-year-old men. The probability of a baby suffering cot death, quoted above, would have been estimated by reference to a large number of babies at risk and then observing how many succumbed. Frequency probabilities are probabilities of *events*.

The second kind of probability is more subjective and is often referred to as a *degree of belief*. Betting odds are an example. Bookmakers arrive at such odds by reflecting on all the factors of which they are aware and making a subjective judgement. Although they may be very experienced and have copious information at their disposal, in the end the number at which they arrive is subjective. There is no absolute standard to which their judgement can be referred to decide whether it is right or wrong. Degrees of belief, like relative frequencies, can also refer to events as, for example, if we consider the possibility that there will be a change of government at the next general election. They can also refer to propositions, such as when we consider the probability that God exists. There are sophisticated ways in which degrees of belief can be elicited, by offering hypothetical bets, for example, and there are also crude subjective judgements arrived at with little thought. For the moment I have lumped all of these together in the second category of *degree of belief*.

MULTIPLYING PROBABILITIES

Probability theory is not so much concerned with estimating individual probabilities as with combining them to determine probabilities of complicated events. For present purposes, by far the most important of these relate to *conjunctions*. If something occurs with probability 0.5 then what is the probability that it will happen ten times in a row if it is repeated? If the chance of a baby dying from cot death is 0.00012, then what is the probability of two babies in the same family dying from the same cause? The answer to this question was given in Sally Clark's trial as $0.00012 \times 0.00012 = 0.0000000144$ or about one in 69 million. What is the justification for multiplying probabilities in this way? The answer is that it is only valid if the events in question are *independent*; that is, if what happens in one case is entirely uninfluenced by the other. When tossing coins this is a plausible assumption because it is difficult to see how the two tosses could be linked physically in any way. But in the case of cot death it is far from obvious that two deaths in the same family would be independent. Common influences, genetical and environmental, are not unusual in families and it would be surprising if events happening to siblings were independent. If events are not independent then the second probability in products such as the above must be modified to take account of what happened to the first baby. This could substantially alter the answer. The same considerations apply if there are more than two events. Multiplying several very small numbers produces a product which can be very small indeed. It is on such calculations that many of the more startling conclusions in the science–religion debate rest. The validity of the implicit assumption of independence is seldom commented on or even noticed!

In a section entitled 'Universes galore: the problem of duplicate beings', Paul Davies (2006, pp. 201–3) discusses calculations purporting to show that in a large enough universe, or collection of universes, all sorts of unlikely things are bound to happen sooner or later. Thus one would certainly find another planet, exactly like the earth, on which lives an individual exactly like oneself in every detail. This would pose obvious problems about personal identity. Davies begins with coin tossing showing that, if you toss a coin sufficiently often, such extremely unlikely sequences as a thousand heads in a row will certainly occur. He then reports some calculations by the cosmologist Max Tegmark about how far, for example, one would expect to travel through the universe before finding an exact copy of oneself. The distance is enormous (about 10^{29} metres), but we are told that 'Weird though these conclusions may seem, they follow incontrovertibly from the logic of simple statistics and probability theory.' Such conclusions may be weird but they certainly do not follow from 'the logic of simple statistics and probability theory'. In addition to assumptions about the uniformity of the laws of physics and suchlike, we may be sure that, lurking somewhere, there will be assumptions of independence such as those used in the introduction about coin tossing; these do not come as part of the probability package. They have to be imported as part of what is assumed to be 'given'. Probability theory must not be saddled with responsibility for every weird conclusion drawn from its use. The weirdness almost always lies in the assumptions, not the logic. If one has a sufficiently lively imagination it is possible to make all sorts of assumptions for whose truth one has no evidence whatsoever. Conclusions are no better than the assumptions on which they are based. Science fiction thrives on improbable scenarios, and though readers may enjoy the thrill of being transported into realms of near-fantasy on the wings of elementary statistics, they should be on their guard against being taken for a ride by calculations based on a grain of fact and a mountain of guesswork.

The probability of life emerging on earth or the remarkable coincidence of the parameter values on which the evolution of an inhabitable world depends are other examples we shall meet later. Only rarely, with such artificial events as coin tossing, is one likely to be confident in assuming independence.[3]

CONDITIONAL PROBABILITIES

One of the most common traps for the unwary is in failing to notice that all probabilities are *conditional probabilities*. This means that the numerical value of a probability depends on what we already know about the event or proposition in question. The reason that this rather obvious fact is so often overlooked is that the conditioning circumstances are often 'understood'. For example, we talk about the probability that a coin falls heads without feeling the need to specify all the circumstances of the toss because they do not, as far as we can see, affect the outcome. The probability arguments used in physics are often of this kind and the notation used makes no mention of such irrelevant factors but many of the fallacious probability arguments we shall meet in the next chapter fail by overlooking this elementary fact. If a doctor is trying to diagnose the cause of a patient's symptoms, then it is obvious that the probability of a particular diagnosis depends on how much the doctor knows about the patient. Tests will be carried out to clarify the position and as the results come in so will the probability of the possible diagnoses change. In order to make this explicit, and to keep track of what is going on, it is necessary to introduce a notation which will incorporate all the relevant information. All that the reader will have to cope with is a new notation which is no more

---

[3] The late William Kruskal, a Chicago statistician, also had a serious interest in theology and he clearly recognised the pitfalls of ignoring independence. His paper on 'Miracles and statistics' was subtitled 'The casual assumption of independence' (Kruskal 1988).

than a convenient shorthand. Probabilities will be written as follows:

P(A GIVEN B).

The P stands for probability and the A and the B within the brackets denote the two things we need to know to specify a probability. A is the thing (event or proposition) in whose probability we are interested. The B specifies the conditions under which the probability is to be calculated (what is assumed to be known). Thus, for example, we might be interested in the probability that someone who is now forty years of age would still be alive at seventy. This obviously depends on a variety of things about the person. So B might, in this case, be male or female and the numerical value of the probability will depend on which is chosen. Again, we might be interested in the effect of smoking on the probability. Thus we would want to calculate,

P(40-year-old alive at 70 GIVEN that they are male and a smoker).

The conditioning event in this case is a compound event depending on two things about the subject. In general, the conditioning event may be as complicated as is necessary. We would not ordinarily include in B anything which we knew to have no effect on the probability. Thus although we could write

P(head GIVEN six heads in a row already)

we would not usually do so because, in independent tosses, it is generally recognised that the probability is unaffected by what has happened prior to that point. The important point to notice and to bear in mind, especially when we come to consider particular probabilities in the next and later chapters, is that every probability is calculated under specific assumptions about what is given. To ignore this elementary fact can be to land in serious trouble.

Another common error is to overlook the fact that the order in which A and B are written matters and can be crucial. Formally, it is *not* generally true that P(A GIVEN B) is equal to P(B GIVEN A). Treating these two probabilities as the same is often known as the *prosecutor's fallacy* because it sometimes arises in legal cases.[4] Suppose that DNA found at the site of a crime is compared with that obtained from a suspect and that the two match. It may be claimed in court that the probability of such a match occurring with the DNA of a randomly chosen individual in the population is one in several million. The jury is then invited to infer that the accused is guilty because what has happened would be extremely unlikely if the accused were not guilty. Let us express this in the formalism set out above. The probability, which is said to have a value of one in several million, may be written

P(match GIVEN accused not guilty).

If we (incorrectly) reverse the elements inside the brackets we would interpret this as

[4] Examples are not confined to the field where the term was coined. One of the earliest errors of this kind was due to Hoyle and Wickramasinghe and before them, Le Comte du Noüy, who used it to prove that God exists. In the next chapter we shall see that because some occurrence in the physical world has an incredibly small probability GIVEN the chance hypothesis, it cannot be assumed that this was the probability of the chance hypothesis GIVEN the occurrence and hence the probability of its complement (that God was the cause) was very close to one. More details will be found in in Bartholomew (1984), see especially chapter 3. Overman (1997) mentions these and other examples and falls into the same trap.

P(accused not guilty GIVEN match).

As this is negligibly small, the complementary event, that the accused is guilty, seems overwhelmingly probable. The reasoning is fallacious but an over-zealous prosecutor and lay jury can easily be persuaded that it is sound. It is true that there are circumstances in which these two probabilities might happen to coincide but this is not usually the case. I shall not go further into the technicalities of why the reasoning behind the prosecutor's fallacy is mistaken but one can get an indication that something might be wrong by the following consideration. Intuitively, one would expect that the judgement of guilt ought to depend also on the probability of being not guilty in the absence of any evidence at all – but here this does not come into the picture.

The prosecutor's fallacy often arises when very small probabilities are involved, as in the above example. Very small probabilities also play a key role in judging whether biologically – and theologically – interesting events could have occurred. It is, therefore, particularly important to be alert for examples of A and B being inadvertently reversed.

CHAPTER 6

What can very small probabilities tell us?

It is often claimed that such things as the origin of life on earth and the coincidental values of cosmic constants are so improbable on the 'chance hypothesis' that they must point to divine action. The correctness of this conclusion depends on both correct calculation and valid forms of inference. Most examples fail on both counts. Two common mistakes in making probability calculations were identified in chapter five. In this chapter I explain and illustrate the three main approaches to inference: significance testing, likelihood inference and Bayesian inference.

WHAT IS THE ARGUMENT ABOUT?

It is tempting to argue that if something has a very small probability we can behave as if it were impossible. When we say 'small' here we are usually thinking of something that is extremely small, like one in several millions. We run our lives on this principle. Every day we are subject to a multitude of tiny risks that we habitually ignore. We *might* be struck by a meteorite, contract the ebola virus, find a two-headed coin or forget our name. Life would come to a halt if we paused to enumerate, much less prepare, for such possibilities. Even the most determined hypochondriac would be hard put to identify all the risks we run in matters of health. Surely, then, we can safely reject the possibility of anything happening if the probability is extremely small? Actually the position is more subtle and we must be a little more careful about what we are saying.

Every time we play a card game such as bridge we are dealt a hand of thirteen cards. The probability of being given any particular hand is extremely small – about one in 6.4×10^{11} and yet it undoubtedly has happened and we have the hand before us as evidence of the fact. That very small probability does not, however, give us grounds for doubting that the pack was properly shuffled. There are other situations like this, where *everything* that could possibly happen has a very small probability, so we know in advance that what will have happened will be extremely unlikely. It is then *certain* that a very rare event will occur! Does this mean that there is never any ground to dismiss any hypothesis – however absurd?

It is clear that there is a problem of inference under uncertainty here with which we must get to grips. Before plunging into this major topic, I pause to mention a few examples where very small probabilities have been used as evidence for the divine hand. In the more spectacular cases these have been held to show that there must have been a Designer of the universe. It is important, therefore, that we get our thinking straight on such matters. Actually there is a prior question which arises concerning all of these examples; not only must we ask whether the inferences based on them are legitimate, but whether the probabilities have been correctly calculated in the first place. Most fail on both counts.

EXAMPLES OF VERY SMALL PROBABILITIES

A common line of argument, employed by many Christian apologists, is that if we can find some event in nature which, on chance alone, would have an extremely small probability, then chance can be ruled out as an explanation. The usual alternative envisaged is that the event must have been caused by the deity, or someone (or thing) operating in that role. It is not hard to find such events and, even if their probabilities cannot always be calculated exactly, enough can usually be said about them to establish that they are, indeed, very small. Some of these examples are longstanding and have been sufficiently discredited not to need extended treatment here. A number were given in *God of Chance* (Bartholomew 1984), and Overman (1997) reports five such calculations in his section 3.7.

The earliest example appears to have been John Arbuthnot's discovery that male births exceeded female births in London parishes in each of the eighty-two years from 1628 to 1710. Le Comte du Noüy looked at the more fundamental question of the origin of life and considered the probability that it might have happened by the fortuitous coming together of proteins to form an enzyme necessary for life to exist. He quoted the extraordinarily small probability of $2.02 \times (1/10)^{321}$ (Hick 1970, p. 15). This is an example of a mistake in the calculation. It falls into the common error of supposing that enzymes are formed by a kind of random shaking together of their constituent parts. Hick, appealing to Matson (1965), recognised this and rejected the calculation. No biologist supposes that enzymes were formed in such a way and the calculation is not so much wrong as irrelevant. In effect this is another attempt to use the multiplication rule for independent probabilities, when the probabilities involved are not independent. Ayala (2003)[1] makes this point very clearly (see, especially page 20). He returns to

---

[1] This paper by Ayala is an excellent review of the 'argument from design' by a biologist and philosopher. Apart from dealing with the matter of small probabilities the paper contains much that is relevant to the debate on Intelligent Design, which is treated in the following chapter.

this topic in his most recent book (Ayala 2007). In chapter 8, especially on pages 153 and 154, Ayala points out that the assumptions on which such calculations rest are simply wrong. There would, perhaps, be no point in relating these episodes were it not for the fact that the same error is continually being repeated. In his attempt to show that life was 'beyond the reach of chance', Denton (1985) uses essentially the same argument, taking as an illustration the chance of forming real words by the random assemblage of letters. This example has been discussed in more detail in Bartholomew (1996, pp. 173ff.)[2] but it suffices here to say that Denton's calculation is based on an oversimplified model of what actually happens in nature. In addition to Overman's examples mentioned earlier, Hoyle and Wickramasinghe (1981) made the same kind of error and this deserves special mention for two reasons. First, it appears to have given currency to a vivid picture which helps to grasp the extreme improbabilities of the events in question. Hoyle compares the chance to that of a Boeing 747 being assembled by a gale of wind blowing through a scrap yard.[3] This analogy is widely quoted, misquoted (one example refers to a Rolls Royce!) and misinterpreted as any web search will quickly show. It was in Hoyle's book *The Intelligent Universe* (1983, p. 19) where it appears as:

[2] Ayala (2007) also uses the analogy of words formed by typing at random in a section beginning on page 61, headed 'A monkey's tale'. He then proposes a more realistic version on page 62, which is based on the same idea as the example referred to here. This example first appeared in Bartholomew (1988).

[3] None of the quotations I have tracked down give a precise reference to its origin. However, on the web site http://home.wxs.nl/~gkorthof/kortho46a.htm Gert Kortof says that the statement was first made in a radio lecture given by Hoyle in 1982.

A junkyard contains all the bits and pieces of a Boeing-747, dismembered and in disarray. A whirlwind happens to blow through the yard. What is the chance that after its passage a fully assembled 747, ready to fly, will be found standing there?

Inevitably, for such a famous utterance, there is even doubt as to whether Hoyle was the true originator. Whether or not the probability Hoyle and Wickramasinghe calculated bears any relation at all to the construction of an enzyme is very doubtful, but its vividness certainly puts very small probabilities in perspective.

Secondly, their argument is a flagrant example of the prosecutor's fallacy, or the illegitimate reversal of the A and B in the probability notation above (see p. 75). For what Hoyle and Wickramasinghe actually claimed to have calculated was the probability that life would have arisen GIVEN that chance alone was operating. This is not the same as the probability of chance GIVEN the occurrence of life, which is how they interpret it. That being so impossibly small, Hoyle and Wickramasinghe wrongly deduce that the only alternative hypothesis they can see (God) must be virtually certain.

In more recent examples the errors in calculation are sometimes less obvious, usually because no explicit calculation is made. Here I mention two very different examples. The first lies behind Stephen J. Gould's claim that if the film of evolution were to be rerun, its path would be very different and we, in particular, would not be here to observe the fact! We shall meet this example again in chapter 11 where its full significance will become clear. Here I concentrate on it as an example of a fallacious probability argument. Gould visualises evolution as a tree of life. At its root are the original, primitive, forms from which all life sprang. At each point in time at which some differentiating event occurs, the tree branches. Evolution could have taken this path or that. The tree thus represents all possible paths which evolution could have taken. The present form of the world is just one of many end-points at the very tip of one of the branches. All of the other tips are possible end-points which happen not to have been actualised. In reality our picture of a tree is limiting, in the sense that there are vastly more branches than on any tree we are familiar with. The probability of ending up at our particular tip is the probability of taking our particular route through the tree. At each junction there are several options, each with an associated probability. These probabilities need not be individually small but, if successive choices are assumed to be independent, the total probability will be the product of all the constituent probabilities along the path. Given the enormous number of these intermediate stages, the product is bound to be very small indeed. From this Gould's conclusion follows – or does it? His assumption was tacit but wrong. It is highly likely that the branch taken at any stage will depend on some of the choices that have been made earlier and that other environmental factors will be called into play to modify the simplistic scheme on which the independence hypothesis depends. In fact there is no need to rely on speculation on this point. As we shall see in chapter 11, there is strong empirical evidence that the number of possible paths is very much smaller than Gould allows. In that chapter we shall look at Simon Conway Morris' work on convergence in evolution. This draws on the evidence of the Burgess shales of British Columbia just as Gould's work did. It shows how facile it can be to make the casual assumption of independence and then to build so much on it.

My second example is quite different and is usually discussed in relation to the *anthropic principle*.[4] It relates to the

---

[4] The anthropic principle was discussed from a statistical point of view in Bartholomew (1988, pp. 140ff.), where some references are given. It has

remarkable coincidence in the basic parameters of the universe which seem to have been necessary for a world of sentient beings to come into existence and to observe the fact! Any list of such coincidences will include the following:

(i) If the strong nuclear force which holds together the particles in the nucleus of the atom were weaker by more than 2 per cent, the nucleus would not hold together and this would leave hydrogen as the only element. If the force were more than 1 per cent stronger, hydrogen would be rare, as would elements heavier than iron.

(ii) The force of electromagnetism must be 10^{40} times stronger than the force of gravity in order for life as we know it to exist.

(iii) If the nuclear weak force were slightly larger there would have been little or no helium produced by the big bang.

In combination, these and other coincidences of the same kind add up to a remarkable collection of surprising facts about the universe. It thus turns out that the values of many basic constants have to be within very narrow limits for anything like our world to have emerged. Probabilities are brought into the picture by claiming that the chance of such coincidences is so remote that we must assume that the values were 'fixed' and that the only way they could have been fixed was for there to be a supreme being who would have done it. This is seen by many

been used as an argument for God's existence by, for example, Montefiore (1985), where the list of coincidences which he gives in chapter 3 is somewhat different from that given here. Other discussions will be found in Holder (2004) and Dowe (2005). Sharpe and Walgate (2002) think that the principle has been poorly framed and they wish to replace it by saying 'that the universe must be as creative and fruitful as possible'. The most recent discussion in a much broader context is in Davies (2006). An illuminating but non-probabilistic discussion will be found in 'Where is natural theology today?' by John Polkinghorne (2006). Dembski (1999, p. 11) regards these coincidences as evidence of Intelligent Design.

to be one of the most powerful arguments for the existence of God – and so it may be – but the probabilistic grounds for that conclusion are decidedly shaky. The argument consists of two parts. First it is argued that the probability of any one parameter falling within the desired range must be infinitesimally small. Secondly, the probability of them *all* falling within their respective ranges, obtained by multiplying these very small probabilities together, is fantastically small. Since this rules out the possibility of chance, the only option remaining is that the values were deliberately fixed.

Let us examine each step in the argument, beginning with the second. There is no obvious reason for supposing that the selections were made independently. Indeed, if there were some deeper model showing how things came about, it could well be that some of the parameter values were constrained by others or even determined by them. The total probability might then be much larger than the assumption of independence suggests. But surely something might be salvaged because there is so little freedom in the determination of each parameter treated individually; this brings us back to the first point. Why do we suppose it to be so unlikely that a given parameter will fall in the required small interval? It seems to depend on our intuition that, in some sense, all possible values are *equally likely*. This is an application of the *principle of insufficient reason* which says that, in the absence of any reason for preferring one value over another, all should be treated as equally likely. The application of this principle is fraught with problems and it does not take much ingenuity to construct examples which show its absurdity. The best we can do is to say that *if it were the case that all values were equally likely* and also *if the parameters were independent*, THEN it would follow that the creation 'set-up', was, indeed extremely unlikely if nothing but chance were operating. But there is no evidence whatsoever for either of these assumptions – they merely refer to our state of mind on the issues and not to any objective feature of the real world.

It should now be clear that the various very small probabilities on offer in support of theistic conclusions must be treated with some reserve. But that still leaves us with the question of how we should interpret them if they were valid.

INFERENCE FROM VERY SMALL PROBABILITIES

The question is: if some event, or proposition, can be validly shown to have an exceedingly small probability on the chance hypothesis (whatever that might turn out to mean) on what grounds can we reject the hypothesis? As noted already, this is the question which Dembski seeks to answer in his support of Intelligent Design but, as Dembski himself recognises, this is not a new issue but one older even than modern statistics. We shall look at Dembski's arguments in the following chapter but first, and without introducing the complications of the contemporary debate, consider the problem from first principles.

First, I reiterate the point made in relation to dealing a bridge hand selected from a pack of cards, namely that a small probability is not, of itself, a sufficient reason for rejecting the chance hypothesis. Secondly, there are three broad approaches, intimately related to one another, which are in current use among statisticians. I take each in turn.

I start with significance testing, which is part of the everyday statistical routine – widely used and misused. If we wish to eliminate chance from the list of possible explanations, we need to know what outcomes are possible if chance *is* the explanation. In most cases all possible outcomes *could* occur, so chance cannot be ruled out with certainty. The best we can hope for is to be nearly certain in our decision. The basic idea of significance testing is to divide the possible outcomes into two groups. For outcomes in one group we shall reject the chance hypothesis; the rest will be regarded as consistent with chance. If the outcomes in the rejection set taken together have a very small probability, we shall be faced with only two possibilities (a) the chance hypothesis is false or (b) a very rare event has occurred. Given that very rare events are, as their name indicates, very rare, one may legitimately prefer (a). If we make a practice of rejecting the chance hypothesis whenever the outcome falls in the rejection set we shall only be wrong on a very small proportion of occasions when that is, indeed, the case. The probability that we shall be wrong in accepting the chance hypothesis on any particular occasion is therefore very small.

In essence this is how a significance test works, but it leaves open two questions (i) how small is 'small' and (ii) how should we construct the rejection set?

This brings us to a second version of the approach, often associated with the names of Neyman and Pearson, for which I begin with the second of the above questions. Let us imagine allocating possible outcomes to the rejection set one by one. The best candidate for inclusion, surely, is the one which is furthest from what the chance hypothesis predicts, or alternatively perhaps, the least likely. In practice, these are often the same thing. In fact, in this case, we may define 'distance' by probability, regarding those with smaller probability as further away. We can then go on adding to the rejection set, starting with the least probable and working our way up the probability scale. But when should we stop? This is where the answer to the first question comes in. We want to keep the overall probability small, so we must stop before the collective probability of the rejection set becomes too large. But there are considerations which pull us in opposite directions. On the one hand we want to keep the probability very small and this argues for stopping early. On the other, if we make the net too small, we reduce the chance of catching anything. For the moment I shall describe the usual way that statisticians take out of this dilemma. This says that there are two sorts of mistakes we can make and that we should try to make the chance of both as small as possible. The first kind of mistake is to reject the chance hypothesis when it is actually true. The probability of this happening is called the 'size of the rejection region'. The second kind of error is failing to reject the chance hypothesis when we ought to – that is when some other hypothesis is true. This argues for making the rejection region as large as possible, because that way we make the net larger and so increase our chance of catching the alternative. The reader may already be familiar with these two kinds of error under the guise of false positives and false negatives. If, for example, a cervical smear test yields a positive result when the patient is free of disease, this is said to be a *false positive* (error of Type I). If a patient with the condition yields a negative result that is described as a *false negative* (error of Type II). In both contexts any rule for deciding must be a compromise.

Let us see how this works out in two particular cases. As noted earlier, John Arbuthnot claimed to have found evidence for the divine hand in the records of births in London parishes. In the eighty-two years from 1629 to 1710 there was an excess of male births over female births in every one of those years. If the chances of male and female births were equal (as Arbuthnot supposed they would be if chance ruled), one would have expected an excess of males in about half of the years and an excess of females in the rest. This is not what he found. In fact the deviation from the 'chance' expectation was so great as to be incredible if sex was determined as if by the toss of a coin. Can we eliminate chance as an explanation in this case? Arbuthnot thought we could and he based his case on the extremely small probability of achieving such an extreme departure from expectation if 'chance ruled'. It would have been remarkably prescient of him to have also considered the probability of failing to detect a departure if one really existed, but this was implicit in his deciding to include those cases where the proportion of males, or females, was very different from equality.

My second example is included partly because it has figured prominently in Dembski's own writings, to which we come in the next chapter. It concerns the case of Nicolas Caputo, the one time county clerk of New Jersey county in the United States. It is a well-established empirical fact that the position on the ballot paper in an election influences the chance of success. The position is determined by the clerk and, in this case, the Democrats had first position on forty out of forty-one occasions. Naturally, the Republicans suspected foul play and they filed a suit against Caputo in the New Jersey Supreme Court. If the clerk had determined the position at random, the two parties should have headed the list on about half of the occasions. To have forty of forty-one outcomes in favour of the party which Caputo supported seemed highly suspicious. Such an outcome is certainly highly improbable on the 'chance' hypothesis – that Caputo allocated the top position at random. But what is the logic behind our natural inclination to infer that Caputo cheated? It cannot be simply that the observed outcome – forty Democrats and only one Republican is highly improbable – because that outcome has exactly the same probability as any other outcome. For example, twenty Democrats and twenty-one Republicans occurring alternately in the sequence would have had exactly the same probability, even though it represents the nearest to a fifty:fifty split that one could have. Our intuition tells us that it is the combination of Caputo's known preference with the extreme outcome that counts. According to the testing procedure outlined above, we would choose the critical set so that it has both a very small probability *and* includes those outcomes which are furthest from what we would expect on the chance hypothesis that Caputo played by the rules.

INFERENCE TO THE BEST EXPLANATION

This is a second way of approaching the inference problem.[5] I borrow the terminology from philosophers and only bring statisticians into the picture at a later stage. This is a deliberate choice in order to bring out a fundamental idea which is easily lost if I talk, at this stage, about the statistical term 'likelihood'. *Inference to the Best Explanation* is the title of a book by philosopher Peter Lipton (Lipton 2004), whose ideas have been seen as applicable to theology. The term *inference to*

[5] There is an interesting parallel between developments in matters of inference in philosophy and statistics which has not, as far as I am aware, been remarked upon before. With very few exceptions, these have been entirely independent. The basic idea of significance testing is that hypotheses are rejected – not accepted. A statistical hypothesis remains tenable until there is sufficient evidence to reject it. In philosophy Karl Popper's idea of falsification is essentially the same. The likelihood principle, introduced into statistics in the 1920s, has surfaced in philosophy as inference to the best explanation, which is discussed in this section. Bayesian inference, to which we come later in the chapter, is becoming popular in philosophy after the pioneering work of Richard Swinburne set out in *The Existence of God* (Swinburne 2004). In his earlier work he referred to Bayesian inference as Confirmation Theory. A recent example is Bovens and Hartmann (2003). One important difference is that statisticians give probability a more exclusive place. For example, the judgement of what is 'best' in Lipton's work does not have to be based on probability. There is an interesting field of cross-disciplinary research awaiting someone here.

the best explanation is almost self-explanatory. It is not usually difficult to conceive of several explanations for what we see in the world around us. The question then is: which explanation is the best? What do we mean by 'best' and how can we decide which is best? This is where probability considerations come in.

Suppose we return home one day and find a small parcel on the doorstep. We ask ourselves how it came to be there. One explanation is that a friend called and, since we were not at home, left the parcel for us to find. An alternative is that a carrier pigeon left it. Suppose, for the moment, that no other explanation suggests itself to us. Which is the better explanation? The 'friend' hypothesis is quite plausible and possible; the pigeon hypothesis much less so. Although it is just possible, it seems scarcely credible that a pigeon should be dispatched with the parcel and choose to deposit it on a strange doorstep. The appearance of the parcel seems much more likely on the 'friend' rather than the 'pigeon' explanation. In coming to this conclusion we have, informally, assessed the probability of a parcel's being deposited by a friend or a pigeon and concluded that the friend provides the better explanation. Let us set out the logic behind this conclusion more carefully. There are two probabilities involved. First there is the probability that a friend decided to deliver a parcel and, finding us out, deposited it on the doorstep. The event we observe is the parcel. We might write the relevant probability as P(parcel GIVEN friend). Similarly on the pigeon hypothesis there is P(parcel GIVEN pigeon). We judge the former to be much higher than the latter and hence prefer the friend explanation because it makes what has happened much more likely. In that sense the friend is the better explanation.

The procedure behind this method of inference is to take each possible explanation in turn, to estimate how probable each makes what we have observed and then to opt for the one which maximises that probability. In essence that is how inference to the best explanation works in this context.

There is one very important point to notice about all of this – it merely helps us to choose between some given explanations. There may be many other possible explanations which we have not thought of. It does not, therefore, give us the 'best' explanation in any absolute sense but merely the best among the explanations on offer. Secondly, notice that the probabilities involved do not have to be large. It need not be very likely that our friend would have left a parcel on our doorstep – indeed, it may be a very unlikely thing to have happened. All that matters is that it should be much more likely than that a pigeon put it there. The *relative* probabilities are what count. If the probabilities are so very low, there would be a powerful incentive to look for other explanations which would make what has happened more likely, but that is a separate issue.

The debate between science and religion, when pursued as objectively as possible, involves choosing between explanations. Two explanations for life in the universe are (a) that God created it in some fashion and (b) that it resulted, ultimately, from the random nature of collisions of atoms (as it has been put). If we could calculate the probabilities involved, it might be possible to use the principle to decide between them. Dowe (2005),[6] for example, uses the principle in relation to the fine tuning of the universe and the evidence which that provides for theism.

This principle of choice has been one of the key ideas in the theory of statistics for almost a century. It is also particularly associated with the name of Sir Ronald Fisher and especially with a major publication in 1922, although the idea went back

[6] The discussion begins on p. 154 in the section on God as an explanation.

to 1912 (see Fisher Box (1978), especially number 18 in the list of *Collected Papers*). In statistics the principle is known as *maximum likelihood*, where the term likelihood is used in a special sense which I shall explain below. One advantage of drawing attention to this connection is that it facilitates the elucidation of the limitations of the principle.

Because the ideas involved are so important for inference under uncertainty, we shall look at a further example constructed purely for illustrative purposes. This is still to oversimplify, but it enables me to repeat the essential points in a slightly more general form.

Suppose tomorrow's weather can be classified as Sunny (S), Fair (F) or Cloudy (C) and that nothing else is possible. We are interested in explaining a day's weather in terms of the barometric pressure on the previous day. Weather forecasters are interested in such questions though they, of course, would take many more factors into account. To keep this as simple as possible, let us merely suppose the pressure to be recorded, at an agreed time, as either above (A) or below (B) 1,000 mm of mercury, which is near the middle of the normal range of barometric pressure. The forecaster is able to estimate probabilities for tomorrow's weather outcomes *given* today's barometric pressure. We write these probabilities as, for example, P(S GIVEN B). This is shorthand for the probability that it is sunny (S) tomorrow, GIVEN that today's barometric pressure is below 1,000 mm (B). The probabilities available to the forecaster can be set out systematically in a table, as follows.

P(S GIVEN A)	P(F GIVEN A)	P(C GIVEN A)
P(S GIVEN B)	P(F GIVEN B)	P(C GIVEN B)

In the first row are the probabilities of tomorrow's weather GIVEN that the barometer is above 1,000 mm today; in the second row are the corresponding values if the pressure is below 1,000 mm today. A forecaster would choose the row representing today's pressure and look for the option which had the largest probability. To make this more definite, suppose the numbers came out as follows.

0.6	0.3	0.1
0.2	0.4	0.4

If the barometer today is above 1,000 mm, the first row is relevant, and since sunny has the highest probability (0.6), this is the most likely outcome. In the contrary case, fair and cloudy have the same probabilities. Both are more likely than sunny but, on this evidence, there is no basis for choosing between them.

However, the real point of introducing this example is to look at the problem the other way round. Imagine that a day has passed and we now know what the weather is like today but have no record of yesterday's pressure. What now is the best explanation we can give of today's weather in terms of the barometric pressure on the day before? In other words, what do we think the pressure on the day before would have been GIVEN that it is sunny, say, today?

To answer this question, the principle of inference to the best explanation says we should look at the columns of the table. In our example there are only two elements to look at: P(S GIVEN A) and P(S GIVEN B) or, in the numerical example, 0.6 and 0.2. The principle says that we should choose A because it gives a higher probability to what has actually happened than does B (0.6 as against 0.2). Similarly, if it was cloudy today we would say that B was the most likely explanation (because 0.4 is bigger than 0.1).

When we look at the probabilities by column we call them *likelihoods*. Choosing the largest value in the column is thus *maximising the likelihood*. In statistics we talk about the principle of maximum likelihood; in philosophy, as inference to the best explanation. For those with some knowledge of mathematics, we use the term *probability function* when we consider the probability as a function of its first argument, when the second – after the GIVEN – is fixed. It is called the *likelihood function* when it is considered as a function of the second argument, with the first fixed.

We can put the matter in yet another way by saying that when we use the probabilities to look forward in time we are making forecasts; when we use them to look backward in time we are making an inference to the best explanation. This prompts a further observation before we leave this example. The rows in the tables always add up to one because all the possible outcomes are included, so one or other *must* happen – that is, the event that one or other happens is certain. The columns do not add up to one because, when looking down a column, it is only the *relative* values of the entries that matter.

Likelihood inference was used in this fashion in Bartholomew (1996) in relation to the existence of God. However, the likelihood principle is part of Bayes' rule which has been used extensively by the philosopher Richard Swinburne and others. The problem with inference to the best explanation is that it takes no account of anything we may know about the hypothesis in advance. For example, in a country where the barometric pressure was almost always high, the fact of knowing that it was a sunny day would not greatly affect our propensity to believe that the pressure had been high the day before. The high prior probability of the pressure being high is not greatly affected by the additional information that it is sunny today, because the two usually go together anyhow. The trouble with comparing likelihoods is that it does not take into account the *relative* probabilities of hypotheses at the outset. This is what the third approach to inference, Bayesian inference, does for us.

BAYESIAN INFERENCE

I shall not go into the technicalities but, instead, give a further example which illustrates how the idea works. Incidentally, this example also demonstrates that if we use likelihood inference on the question of God's existence, it produces a rather odd conclusion. This fact gives added impetus to the search for something better.

The likelihood comparison always makes the 'God hypothesis' more credible than any competing hypothesis. For simplicity, suppose that we wish to compare just two hypotheses: one that there exists a God capable of bringing into being whatever world he wishes, the other that there is no such God. On the first hypothesis it is certain that the world would turn out exactly as God intended and hence the conditional probability of its happening thus must be one. On the alternative hypothesis we may not be able to determine the probability precisely, or even approximately, but it will certainly be less than one, so the 'God hypothesis' is to be preferred. I suspect that few atheists would be converted by this argument and it is pertinent to ask 'why not?' The whole thing smacks of sleight of hand. It sounds like cheating if we contrive a hypothesis with the sole purpose of making what has happened certain. If you begin with the conviction that God's existence is intrinsically unlikely there ought to be some way of making sure that that fact is taken into account. That is exactly what Bayes' theorem does.

Essentially it asks whether the likelihood is sufficient to outweigh any prior odds against the hypothesis. Looked at in another way, it allows us to bring our own feelings, judgements or prior evidence into the equation. Whether or not we should be allowed to do this is an important question.

In order to get a rough idea of how the theorem works let us introduce the idea of a *likelihood ratio*. This is a natural and simple extension of the ideas already introduced. Let us go back to the example about weather forecasting. There we considered two probabilities, $P(S \text{ GIVEN } A)$ and $P(S \text{ GIVEN } B)$. Their numerical values as used in the example were 0.6 and 0.2, respectively. I regarded A as the more likely explanation of the sunny weather, represented by S, because its likelihood was three times ($=0.6/0.2$) the likelihood of the alternative. This was the likelihood ratio for comparing these two hypotheses. The bigger it is, the more strongly inclined are we to accept A as the explanation. However, if we had a justifiable prior preference for B we might ask how strong that preference would have to be to tip the evidence provided by the likelihood ratio the other way. Bayes' theorem tells us that if the prior odds on B were three to one, that would just balance the evidence from the likelihood ratio the other way. The odds would have to be less than three to one for the evidence of the likelihood ratio to carry the day and more than three to one for it to go the other way.

In general we have to compare the likelihood ratio to the prior probability ratio. The product of the two determines which way the conclusion goes.

We shall meet these ideas again in the next chapter when we move on to the stage where Intelligent Design is debated.

CHAPTER 7

Can Intelligent Design be established scientifically?

Intelligent Design has been proposed as a credible scientific alternative to the theory of evolution as an explanation of life on earth. Its justification depends on an extension of Fisherian significance testing developed by William Dembski. It is shown, in this chapter, that there is a fatal flaw in the logic of his method, which involves a circularity. In order to construct a test to detect design and 'eliminate' chance, one has to know how to recognise design in the first place. Dembski's calculation of the probability required to implement the method is also shown to be erroneous.

WHAT IS THE ARGUMENT ABOUT?

Intelligent Design is at the centre of one of the fiercest debates currently taking place in the science and religion field. Its proponents claim that the scientific establishment is set on an atheistic course by refusing to countenance the possibility that the world might contain evidence of design. All they ask is that design should not be arbitrarily ruled out from the start and that nature should be allowed to speak for itself; no special privileges are asked for. The whole debate should then take place within the bounds of science and according to its principles of rationality.

The opponents will have none of this, claiming that Intelligent Design makes no claims that can be tested empirically and, because it cannot be falsified, it is not science. Many see it as crypto-creationism masquerading under the guise of science. They suspect that it is a scientific front for an ideology whose aims are more sinister and which are carefully concealed.

The United States of America is the birthplace and home of Intelligent Design[1] and it is out of the heady mix of a conservative fundamentalism and the threat of religion's trespassing into education that the heat of the debate comes. If evolution is 'only a theory' then why, it is argued, should not other theories get 'equal time' in education? After all, should not children be given the opportunity to make up their own minds on such an important matter and should not their parents have a say in what their children are taught? The proponents of Intelligent Design, it is alleged, wear the clothes of liberals pleading that all sides should be given a fair hearing, whereas, from the other side, the scientific establishment is presented as a group of reactionaries seeking to control what is taught.

---

[1] There is an enormous literature on this topic, much of it highly controversial. It is difficult to select a few articles as background reading. A broad, if uncritical, survey will be found in O'Leary (2004) about a quarter of whose book consists of notes, which reflect the author's wide reading. O'Leary is a journalist who makes no pretence of being an expert in science. Her sense of fairness doubtless leads to her tendency to grant 'equal time' to all sides of an argument. This means that minority viewpoints may appear, to the uninitiated, to carry more weight than they ought. This is true, for example, of the few creationists who have scientific qualifications. The far more numerous members of the scientific establishment are treated with less sympathy. A more academic treatment will be found in Peterson (2002). The journal *Perspectives on Science and Christian Faith* (the journal of the American Scientific Affiliation) has carried two extensive debates that reflect the divisions on the subject among more conservative Christians: volume 54 (2002): 220–63 and volume 56 (2004): 266–98. Much of the technical material is due to William Dembski and this will be referred to in the course of the chapter.

This is not a private fight confined to one country but, since it goes to the heart of what is true in both science and religion, anyone may join in. Those who start as spectators may well see more of the game (to change the metaphor) and so have something to contribute.

The main thesis of the Intelligent Design movement runs counter to the central argument of this book. Here I am arguing that chance in the world should be seen as *within* the providence of God. That is, chance is a necessary and desirable aspect of natural and social processes which greatly enriches the potentialities of the creation. Many, however, including Sproul, Overman and Dembski, see things in exactly the opposite way. To them, belief in the sovereignty of God requires that God be in total control of every detail and that the presence of chance rules out any possibility of design or of a Designer.

To such people, the fact that evolution by natural selection involves chance in a fundamental way appears to rule out the design and purpose without which the whole theistic edifice collapses. Defence of theism thus involves showing that chance is non-existent or, at best, is no more than a description of our ignorance. It is allowed to have no ontological status at all. The Intelligent Design movement is dedicated to showing that the world, as we know it, simply could not have arisen in the way that evolutionary theory claims. This is not to say that evolution by natural selection could not have played some part, but that it could have done so only in a secondary manner. The broad picture could not, they argue, have come about without the involvement of a Designer. In this chapter I shall examine the claim that it can be rigorously demonstrated that chance does not provide a sufficient explanation for what has happened. I shall not be directly concerned with other aspects of Dembski's argument, in particular his claim that information cannot be created by chance.

Essentially, there are two matters to be decided. Is the logic of the argument which, it is claimed, leads to the 'design' conclusion valid and, if it is, are the probability calculations which it requires correct? The logic is effectively that of significance testing outlined in the last chapter. According to the chief theoretician among the proponents of Intelligent Design, William Dembski, the logic is an extension of Sir Ronald Fisher's theory of significance testing. Given Fisher's eminence as a statistician, it is appropriate that his voice should be heard on a matter so close to the core of his subject. Probability calculations also come under the same heading, so I shall examine how Dembski made his calculations.

William Dembski has single-mindedly pursued his goal of establishing Intelligent Design as a credible alternative to evolution in several major books and a host of other articles, books and lectures. This publication trail started with *Design Inference*, in which he set out the basic logic of eliminating chance as an explanation of how things developed. This was followed by *No Free Lunch* and *The Design Revolution* (Dembski 1998, 2002 and 2004).[2] The latter book is subtitled *Answering the Toughest Questions about Intelligent Design* and is, perhaps, the clearest exposition of his basic ideas for the non-technical reader. He has also collaborated with Michael Ruse in editing *Debating Design; From Darwin to DNA* (Dembski and Ruse 2004).

Much of Dembski's argument is highly technical, and well beyond the reach of anyone without a good preparation in mathematics, probability theory and logic. This applies as much to the material written for a general readership as to

---

[2] In *Zygon* 34 (December 1999): 667–75, there was an essay review by Howard J. van Till of both Dembski (1998) and Overman (1997). This is in substantial agreement with the views expressed in this book. Van Till's review was followed by a rejoinder from Paul A. Nelson on pp. 677–82.

the avowedly technical monograph which started it all off (Dembski 1998). In one sense this is highly commendable, because the clarity and rigour which can be attained by this means offers, at least, the prospect of establishing the ideas on a secure scientific foundation, so that the debate can take place within the scientific community.[3] However, this fact poses a serious dilemma for anyone who wishes to engage with him. If the case against Intelligent Design is made at too high a level, it will pass over the heads of many of those who most need to question it. If it is too elementary, it will fail to treat the opposition seriously enough. One must also bear in mind the psychology of the readership. A highly technical treatment can have two opposite effects. On the one hand there is the tendency, on the part of some, to put undue trust in mathematical arguments, thinking that anything which is beyond their reach is also beyond question and almost certainly correct! On the other hand, others may dismiss it instantly as they dismiss all such material, on the grounds that what cannot be expressed in simple everyday language can be ignored as esoteric nonsense. Neither view is correct in this case. The extensive theoretical treatment cannot be so easily dismissed, but neither should it be swallowed whole.

DEMBSKI'S ARGUMENT

To do justice to the subtleties of Dembski's arguments we shall have to take things fairly slowly, but the reader may be

---

[3] The two opposite reactions mentioned in this paragraph will be familiar to anyone who, like the author, has attempted to explain technical – especially mathematical – matters to lay audiences. As noted in the preface, the problem is acute in a book such as this. It often is a case of being 'damned if you do and damned if you don't'.

grateful for a simple statement at the outset of the essence of the situation.

The universe is very large and very old, so there has not been either enough time or enough space for some exceptionally rare events to occur. Roughly speaking, Dembski claims to be able to calculate the probability of the rarest event one could have expected to happen 'by chance' somewhere at some time. It is simply not reasonable to expect any chance event with smaller probability to have occurred at all. Hence if we can find existing biological entities, say, whose probability of formation by chance is less than that critical bound, we can infer that they could not have arisen by chance. Hence they could not have arisen by evolution if that process is driven by chance. Dembski claims that at least one such entity exists – the bacterial flagellum – and that suffices to establish design and, necessarily, a Designer.

It is important to notice that the 'design' conclusion is reached by a process of elimination. According to Dembski there are only two other possible explanations: natural law or chance. Once these are eliminated, logic compels us to fall back on design. It is not often noted that the entity must not only be designed but also brought into being.[4] There must, therefore, be a Maker as well as a Designer. Since natural law can be regarded as a special, but degenerate, case of chance the main thing is to eliminate chance. That is exactly what a Fisherian significance test was designed to do.

THE ELIMINATION OF CHANCE

It can never be possible to eliminate the chance explanation absolutely. The best we can do is to ensure that our probability

---

[4] Howard van Till is an exception. Van Till distinguishes 'the *mind-like* action of *designing* from the *hand-like* action of *actualising* . . . what had first been designed' (2003, p. 128).

of being wrong, when we claim to have eliminated chance, is so small that it can be neglected. Dembski believes that Fisher essentially solved this problem but that his procedure had two gaps which can be closed. When this is done, the way is clear to reach the goal.

I shall begin by recapitulating the basic idea of a test of significance, which starts from the assumption that chance is the explanation and then seeks to demonstrate that what has happened is not consistent with that hypothesis. This time, however, I shall use what is, perhaps, the simplest possible kind of example which still facilitates comparison with what Dembski actually does. We have already seen this example, discussed by John Arbuthnot in chapter 6 on sex determination. Here, as he did, we suppose that the probability of a female birth is exactly 0.5, independently of all other births. In other words, it is just as if sex was determined by coin tossing.

Suppose we have a sample of eight births. The first step is to list all possible outcomes. One possibility is that they all turn out to be male, which we might write as MMMMMMMM; another would be MFMMFFMM, and so on. Altogether there are $2^8 = 256$ possibilities. The next step is to calculate the probability of each outcome. Because of the simple assumptions we have made they all have the same probability of $1/256$. The final step is to construct a rejection set such that any occurrence in that set will lead us to reject the hypothesis. Since we do not wish to reject the hypothesis when it is actually true, the probability of falling in this set should be small. One possible way to do this would be to make our rejection set consist of the two outcomes MMMMMMMM and FFFFFFFF, that is: all male or all female. It seems clear that if one of these goes in, the other should too, because they represent equivalent departures from what one would expect – roughly equal numbers of males and females. The probability associated with this set of two outcomes is 1 / 128, which is not particularly small but it will serve for purposes of illustration if we treat it as 'small'.

If we now adopt the rule that we will reject the chance hypothesis whenever we observe *all males* or *all females* in a set of eight births, we shall wrongly reject the hypothesis one time in 128. Is this a sensible procedure? It certainly ensures that we shall rarely reject the chance hypothesis when it is, in fact, true but that would be the case for any set of two outcomes we might happen to select for the rejection set. What is required, it seems, is a rejection set which has both small probability *and* which 'catches' those outcomes which are indicative of non-randomness or, in Dembski's terminology, design. At this point there is some divergence between Dembski and the traditional statistical approach, as represented by Fisher. It will be instructive to look at these two approaches in turn.

The Fisherian would want to include in the rejection set those outcomes which were 'furthest' from 'chance', in some sense, that is, from what the hypothesis under test predicts. If the probability of a male birth is really 0.5 we would expect around four males in every eight births. An all-male or an all-female outcome would be the most extreme and these would be the first candidates for inclusion in the rejection set. Next would come those with only one male or female, then those with two, and so on. The process would stop when the probability of the rejection set reached the value we had chosen as the small probability that we had set as the risk we were prepared to run of wrongly rejecting the chance hypothesis.

To the end of his life Fisher thought that his procedure just described captured the essence of the way that scientists work, though he strongly objected to the idea of rejection 'rules'. He preferred to quote the *significance level*, which was the size of the smallest set which just included the observed sample. Nevertheless, the distinction is not important for present purposes. Other, perhaps most, statisticians, came to think that more explicit account should be taken of the alternative hypotheses which one was aiming to detect. This was done implicitly in the sex ratio example by constructing the rejection region starting with those samples whose proportion of males, and hence females, was furthest from 0.5. Neyman and Pearson extended the theory to one which bears their name, by arguing that the rejection set should be determined so as to 'catch' those outcomes which were indicative of the alternatives envisaged. Thus, for example, if one were only interested in detecting a tendency for males to be more common, then one would only include outcomes where male births predominated.

It is in choosing the critical region that Dembski's aim is different. He is looking for outcomes which show evidence of design, so his critical region needs to be made up of those outcomes which bear evidence of being non-random. One example of such an outcome would be MFMFMFMF. As male and female births are precisely equal, this outcome would not be allocated to the critical region in the Fisherian approach. This difference directs our attention to the fact that Dembski is actually testing a different hypothesis. In our example, the hypothesis concerned the value of the probability – whether or not it was 0.5. The question we were asking was: are the outcomes consistent with births being determined at random and with equal probability, in other words, just as in coin tossing? Dembski is not concerned with the value of the probability but with the randomness, or otherwise, of the series. It is not altogether clear from his writing whether Dembski has noticed this distinction. He does, however, recognise that the Fisherian scheme needs to be developed in two respects to meet his needs. First he notes that one has to decide what counts as 'small' in fixing the significance level. Dembski claims to have an answer to this question and we shall return to it below. The other point, which is germane to the discussion of how to select the rejection region, is that Dembski wishes to eliminate *all* chance hypotheses not, as in our example, just the one with probability 0.5.

Although it is not totally clear to me how Dembski thinks this should be handled, it is implicit in much of his discussion. Essentially he wishes to include in the rejection set all those outcomes which show unmistakable evidence of design. He calls this property, *specified complexity*. An interesting way of approaching this is through the work of Gregory Chaitin and his way of measuring non-randomness, which I have already discussed in chapter 4. Outcomes exhibiting specified complexity will score highly on a measure of non-randomness and so will be candidates for inclusion. Of all possible outcomes it is known that almost all of them appear to be random, so the proportion which show some pattern form a vanishingly small set which must, inevitably, have small probability. Dembski's approach is slightly different in that he sometimes appears to use 'small probability' as a proxy for 'specified complexity'. This is plausible if one thinks that anything which is designed is bound to be virtually impossible to construct by chance alone and hence must have an exceedingly small probability. Constructing a rejection region by first including outcomes with the smallest probabilities will thus ensure that we only reject the chance hypothesis in favour of something which has specified complexity. However, while it is true that any outcome exhibiting specified complexity will have small probability, the converse is not necessarily true.

All of these ideas are put to the test when we come to consider particular examples, and for Dembski, that means the bacterial flagellum. But first we must return to the question of what is meant by 'small' in this context.

THE UNIVERSAL PROBABILITY BOUND:
HOW SMALL IS SMALL?

According to Dembski, one of the defects of Fisherian significance testing is that it does not say what is meant by 'small' when choosing a significance level. He provides an answer to this question in what he calls the *universal probability bound*. The idea is very simple; the calculation less so. The idea is that the universe is simply not old enough, or big enough, for some events to have materialised anywhere at any time. To put a figure on this requires a calculation of how many events of specified complexity could have occurred. This leads to what Dembski calls the universal probability bound of $1/10^{150}$. I shall not go into the details of his calculation but it depends, for example, on the number of elementary particles in the universe (10^{80}), the rate at which changes can take place, and so on.

It is worth pausing to reflect on the extreme smallness of the probability that I have just been talking about. The number of elementary particles in the universe is, itself, unimaginably large. It is difficult enough to imagine the number of stars in the universe but this difficulty is compounded by the fact that every single star is composed of a vast number of elementary particles. Even when all these are added up we are still many orders of magnitude short of the number 10^{150}. The only point of raising this issue is to turn the spotlight onto the importance of getting the calculation right, if what we are going to do next is compare our calculated probability with some infinitesimally small bound.

The calculation is not straightforward. Although we shall not go into details, there are a number of pitfalls in making such calculations, which Dembski avoids, even though he has to invent a whole new terminology to express what he is about. To begin with, Dembski finds it necessary to introduce the notion of what he calls *probabilistic resources*. This has to do with the fact that there may have been many opportunities at many places for a particular event to occur. So the question we ought to be asking is not whether that event occurs exactly once, but at least once. Another difficulty is that there is simply not always enough information to make an exact calculation. It is sensible, however, to err on the safe side, so Dembski's final answer is not, therefore, an exact figure but a lower bound. This means that the true figure cannot be smaller than this bound, but may be higher. So if, when we make a comparison with another probability, that probability turns out to be smaller than the bound, it will certainly be smaller than the true figure.

There is one more important aspect of Dembski's argument to which we should pay attention. He proposes that the rejection set should consist of the specifically complex outcomes. At this point we run up against the fact that Dembski sometimes appears to regard a rejection set as consisting of one outcome but this is not strictly true. Just as he introduces the idea of probabilistic resources so he introduces *structurally complex* resources. The former allows for the fact that an event which has only a very small probability of occurring at a particular time and place will have a much larger probability if it can occur at many places and times. Similarly, the latter allows for the fact that there may not be just one structurally complex outcome but a number. The probability of observing at least one of them is, therefore, larger than that of exactly one. In effect this means that we have a rejection set consisting of several outcomes just as I supposed in describing the Fisherian significance test.

If we let this collection of specifically complex outcomes constitute the rejection set, we will achieve two objectives at once. First, since the set of specifically complex outcomes is very small, its size (probability) will also be very small, thus meeting the first requirement of a test of significance. Secondly, if we reject the chance hypothesis whenever the outcome falls in this set, we shall never make an error of Type II (false negative). This is because every element in the rejection set is certainly indicative of design, by definition. That is, we may ascribe design to the outcome in such cases without any risk of making a mistake. This is exactly what Dembski wishes to achieve.

A moment's reflection will show that there is something a little odd about this line of reasoning. It says that we should reject the chance hypothesis whenever the outcome exhibits specific complexity. In doing so, we shall certainly be correct if design is, in fact, present and our chance of wrongly rejecting the chance hypothesis will be very small (the size of the rejection set). However, one may legitimately ask why we need all this technical apparatus if we know already that certain outcomes exhibit design. The conclusion is, indeed, a tautology. It says that if something bears unmistakable evidence of design, then it has been designed! The nature of what Dembski is doing, and its absurdity, will be even more obvious when we set it in the context of what he calls 'comparative' methods below. First, I digress to point out the other flaw in Dembski's argument.

THE PROBABILITY OF THE BACTERIAL FLAGELLUM

Although Dembski spends a great deal of time developing a version of Fisherian significance testing designed to eliminate chance, the main application is to one particular case where the theory is not much in evidence. This concerns a remarkable biological structure attached to the bacterium *Escherichia coli*,[5] which drives it in the manner of a propeller. The question is whether this construction could have been assembled by chance or whether its presence must be attributed to design. Dembski makes a rough calculation of the probability that this structure could have come about by chance and arrives at the exceedingly small value of $1/10^{263}$. How on earth, one may wonder, could anyone ever arrive at such a seemingly precise figure? Inevitably there have to be some approximations along the way, but he chooses them so as to err on the safe side. However, there is no need to stay on the details because the whole enterprise is seriously flawed. Howard van Till (2003) has put his finger on the source of the problem. His criticism is that Dembski's probability calculation in no way relates to the way in which the flagellum might conceivably have been formed. Dembski treats it as what van Till calls a *discrete combinatorial object*. Essentially, Dembski counts the number of ways in which the ingredients of the flagellum could be brought together and assembled into a structure. The bland, and false, assumption that all of these structures are equally likely to have arisen then yields the probability.

It is difficult to understand how such an elementary mistake can have been made by someone so mathematically sophisticated. Possibly it stems from confusion about what is meant by 'pure chance'. There are many examples in the literature of similar combinatorial calculations which purport to show that such things as the origin of life must have been exceedingly small. This has already been noted in chapter 6 in relation

---

[5] The case of the bacterial flagellum dominates the literature, almost as though it were the only sufficiently complicated biological organism. Later in this paragraph we come to van Till's discussion of its probability, which was the main purpose of the paper quoted in note 4 above. In strict logic, of course, only one case is needed to establish the conclusion that some things are too complicated to have evolved.

to the work of du Noüy and to Hoyle and Wickramasinghe, among others. As noted in the last chapter, no biologist has ever supposed that such complicated entities can be assembled as a result of some cosmic shuffling system. Indeed, the main point of Dawkins' book *Climbing Mount Improbable* (Dawkins 2006 [1996]) is to demonstrate that complicated structures which it would be virtually impossible to assemble as discrete combinatorial objects could be constructed in a series of small steps which, taken together, might have a much larger probability (see 'Chance in evolution' in chapter 11, below). According to evolutionary theory the growth in complexity would have taken place sequentially over immense periods of time. What is needed is a model of how this might have happened before we can begin to make any meaningful calculations. To produce an argument, as Dembski does, that the flagellum could not have been formed by an 'all-at-once' coming together and random assembly of the ingredients is hardly more than a statement of the blindingly obvious. The inference that Dembski wishes to make thus fails, even if his universal probability is accepted.

THE PARADOX

We now return to the logic of Dembski's argument. Because the fallacy is so fundamental, I shall repeat what was said above but in a slightly different way.

Dembski has always seen his approach as standing squarely in the Fisherian tradition, in which no account needs to be taken of alternative hypotheses. At first sight this seems to be a reasonable position to adopt, because any alternative hypothesis would have to be specified probabilistically and it is the express purpose of the exercise to eliminate *all* chance hypotheses. It is thus somewhat ironic that Dembski's logic can be set out quite simply within the framework of the

Neyman–Pearson approach to inference. The clarity which we gain thereby also serves to underline the essentially tautological character of the formalism.

Let us think of a situation, like the coin-tossing exercise, in which there are very many possible outcomes, each having very small probability (in Dembski's terminology these are complex). Some of these outcomes will be what Dembski calls specifically complex. These outcomes exhibit some kind of pattern which bears the hallmark of design – let us leave aside for the moment the question of whether or not 'pattern' can be adequately defined. The essence of the Neyman–Pearson approach to statistical inference is to choose the rejection set to include those outcomes which are most likely to have arisen under some alternative hypothesis. In this case the alternative is that the outcomes are the result of design. The characteristic of a designed outcome is that it exhibits specified complexity. The rejection set should therefore consist of all those outcomes.

Now let us consider the consequences of what we have done. The likelihood of wrongly rejecting the chance hypothesis is very small because specified outcomes have very small probability. The probability of correctly rejecting the chance hypothesis is one (that is, certain) because all outcomes in the rejection set are certainly the result of design (that is why they were selected). In other words, we have maximised the chance of detecting design when it is present. We thus seem to have a foolproof method of detecting design whose logic has been made clearer by setting it in the Neyman–Pearson framework (which Dembski seems to be hardly aware of). So where is the catch? The problem is that, in order to construct the rejection set, we have to be able to identify those outcomes which are the result of design. If we know that already, why do we need the test in the first place?

One possible response is to say that we only identify design indirectly through the very small probability which it assigns to some outcomes. This would suggest that the rejection region should be formed of those outcomes which have the smallest probabilities and leave, in particular, those which are less than the universal probability bound. In that case we are entitled to ask why we need the formalism at all. If the rule to follow is to reject the chance hypothesis whenever the outcome observed has probability that is so small that it could not have arisen in a universe as large or old as the one we inhabit, is that not a sufficient ground of itself?

DEMBSKI'S CRITICISMS OF COMPARATIVE METHODS

Dembski is highly critical of what he calls comparative methods and his reasons are set out in chapter 33 of Dembski (2004).[6] A comparative method is any method which involves the comparison of the chance hypothesis with an alternative. Such a method involves selecting one from several possibilities and is thus concerned with the relative rather than the absolute credibility of hypotheses. At first sight this is a surprising position to take because there clearly is an alternative in mind – that the complexity we observe is the work of a Designer. However, this alternative clearly has a different status in Dembski's mind, presumably because it is not specified probabilistically. There are three comparative methods in common use which I have already reviewed in chapter 6. The first is the Neyman–Pearson approach, which uses the alternative hypotheses to select the rejection set; the second

---

[6] Dembski mentions (2004, p. 242) a conference at Calvin College in May 2001 on Design Reasoning, at which he spoke. Timothy and Linda McGrew and Robin Collins are reported as putting Bayesian arguments. In particular these critics objected to the notion of *specification*.

is the likelihood method, or inference to the best explanation approach; and the third is the Bayesian method, in which the alternatives have specified prior probabilities. Dembski seems hardly aware of the first two approaches and concentrates his fire on the Bayesian threat. Possibly this is because his own treatment has been challenged from that quarter and this is where much current philosophical interest in uncertain inference lies.

I agree with Dembski's strictures on the use of Bayesian inference in this particular context primarily because the introduction of prior probabilities makes the choice too subjective. In an important sense it begs the question because we have to decide, in advance, the strength of our prior belief in the existence, or not, of a designer. Bayesian inference tells us how our beliefs should be changed by evidence, not how they should be formed in the first place. What Dembski seems to have overlooked is that his method is, in fact, a comparative method and that it can be seen as such by setting it within the framework of the Neyman–Pearson theory as demonstrated in the last section. By viewing it in that way we saw that its tautological character was made clear and hence the whole inferential edifice collapses. Given, in addition, that Dembski's probability calculation is certainly incorrect I conclude that Intelligent Design has not been established scientifically.

IS INTELLIGENT DESIGN SCIENCE?

Much of the debate over Intelligent Design has not been on its statistical underpinning but on the more general question of whether or not it is science. This usually turns on whether it makes empirical statements which can be tested empirically. Although this is true, in my view it approaches the problem from the wrong direction. To make my point it is important to distinguish 'science' from 'scientific method'. Scientific method is the means by which science as a body of knowledge is built up. Dembski has proposed a method by which, he claims, knowledge is validly acquired. The question then is: is this method valid? That is, does it yield verifiable facts about the real world? As I noted at the outset, two questions have to be answered: is the logic sound and is the method correctly applied? To the first question my answer is that the logic is not sound, because the extension proposed to Fisherian significance testing is not adequate in itself and also because almost all statisticians find the original Fisher method incomplete because it ignores alternative hypotheses. To the second question the answer is also negative because the calculation of the key probability is incorrect. The second failure could, in principle, be put right even though the practicalities are almost insurmountable. The first failure seems irredeemable because, once we introduce alternative hypotheses, a circularity in the argument for the construction of the critical set becomes apparent. Dembski's method is not, therefore, a valid scientific method.

CHAPTER 8

Statistical laws

Do statistical laws offer space for God to act in the world? It is first made clear that such laws are not represented primarily by probability distributions but by the basic processes underlying them. One possibility is that the laws are illusory in the sense that God is actually acting in every event in a way that mimics chance. Another possibility is that God acts only occasionally at particularly significant junctures. In this case his purposeful action would be masked by the mass of genuinely random happenings. It is concluded that neither explanation is satisfactory and hence that God's action is more likely to be seen in the behaviour of chance happenings in the aggregate.

ROOM FOR GOD'S ACTION?

From this point in the book onwards, chance plays a positive role. It is no longer to be seen as a threat to theistic belief which has to be ruled out, but as part of the richness of the creation. In the present chapter I investigate whether so-called statistical laws, which involve a chance element, offer a degree of flexibility which might create the space for God to act without disturbing the general lawfulness of the world. Having dipped our toes into the water, so to speak, we shall move on in the following chapter, to see whether the ideas carry over onto the broader canvas of the quantum world.

God's possible action in the world is one of the most intriguing and difficult issues for theologians who wish to pay due regard to the way the world actually works. The subject has spawned an enormous literature and it was singled out for detailed study in the Divine Action Project (DAP), sponsored jointly by the Vatican Observatory and the Center for Theology and Natural Sciences, Berkeley.[1] This programme brought together many of the most distinguished workers in the field. It also included a series of conferences and five main publications in the period 1988–2003. Its impact has been assessed by Wesley Wildman in two articles in *Theology and Science* (Wildman 2004 and 2005). The first paper provides a summary and assessment of the project. The second deals with a number of responses to the first article by several key figures in the debate.

If the world were completely deterministic there would be a rigidity which would require God to break laws in order to change things. Statistical laws have been seen to offer theologians a lifeline by providing room for God to manoeuvre without challenging the overall lawfulness of the creation. This seductive avenue is fraught with hazards and in this chapter I shall begin to show why, by considering the simplest kind of example. We begin with the question of how God might *keep* a statistical law – because the answer is far from obvious. Only then shall we see what might be involved

---

[1] This volume is perhaps the central reference in this field, including, as it does, chapters by most of the major contributors to this subject (Russell *et al.* 1995). Each of the chapters in Sections III and IV deals directly and authoritatively with different aspects of the problem. The last chapter, by George Ellis, is not only of interest in its own right, but also because he brings together, and comments on, many of the earlier contributions, especially that of Nancey Murphy.

in breaking one or, more to the point, how one might take deliberate action within such a law.

WHAT, EXACTLY, IS A STATISTICAL LAW?

This is the crucial question which must be answered before the main issue can be addressed. At first sight it all seems quite simple. Ordinary, non-statistical laws are concerned with things which specify unchanging features of the world. Ohm's law, the law of gravity and Newton's laws of motion are all familiar to us. For example, the current flowing through a resistor is proportional to the drop in potential between the ends. In all these cases the relationship seems to be the same everywhere at all times. However, when referring to statistical laws, we are often thinking of a single quantity – not the relationship between quantities. In the universe there are the so-called constants of nature, such as the speed of light, which are all part of what we mean when we speak of the lawfulness of the world. A statistical law is then generally thought of as replacing the fixed quantity by a probability distribution. This distribution specifies how the quantity *varies* by telling us how frequently, relatively speaking, each different outcome occurs. Looking at it theologically, one might suppose that God chooses the *form* of the distribution but not necessarily the particular values which occur and which, collectively, make up that distribution. Keeping such a law might seem to be a very straightforward matter. All one has to do is to continually check that values are occurring with the correct relative frequencies. If any are not, then all that is necessary is to top up those values which are under-represented and to curtail those that are occurring too frequently. That is evidently what Florence Nightingale thought, as the following episode shows, but she was wrong!

FLORENCE NIGHTINGALE AND 'ILL-DRIVEN CABS'

Muddled thinking about the theology of statistical laws is provided by an example relating to Florence Nightingale. She was much more than 'the lady of the lamp' in the Crimea. She had a lively interest in statistics and although failing in her attempt to persuade the University of Oxford to establish a chair in the subject, she bombarded the powerful of her day with the figures with which she hoped to persuade them to make reforms. Her interests were extremely practical. She thought that there were social laws and that their discovery would enable them to be used to achieve good ends. If God had made society lawful, she reasoned, it was up to us to discover those laws and to use that knowledge for the betterment of society. She wrote:

> . . . of the number of careless women to be crushed in a given quarter under the wheels of ill-driven cabs: were the number not made up on the last days of the Quarter, we await (not with coolness, let us hope) the inexorable law of Fate which – always supposing the state of Society not to be changed always fills up its quota. (Diamond and Stone 1981, p. 77;[2] see also p. 337 of Part III)

---

[2] The article by Diamond and Stone (1981) was entitled 'Nightingale on Quetelet' and it was divided into three parts: I The Passionate Statistician, II The Marginalia, and III Essay in Memoriam. Adolphe Quetelet was the Astronomer Royal of Belgium who wrote a two-volume work entitled *Physique Sociale*, published in 1869. This work was based on the idea that society could be studied scientifically in an analogous way to that in which the physical world was studied in physics. Florence Nightingale was passionately concerned with the quantitative study of society and, for her, Quetelet's work was vital. The article is based on Nightingale's copy of *Physique Sociale*, presented to her by the author in 1872, and on a bundle of her manuscripts tied up with the book in the British Library Department of Manuscripts. The copy of the book contained extensive marginal notes.

The 'inexorable law of Fate' was, of course, essentially an example of a statistical law which, in this instance, governed the number of young ladies falling under cab wheels. In fact, it is very much the same as the archetypal statistical law governing the outcomes of coin tossing, to which we shall come shortly: the correspondence is not exact but near enough for present purposes. According to the law, the expected number of such deaths in a quarter should then be approximately constant from one quarter to another. This is the aspect of the law which caught Florence Nightingale's attention and which could be found repeated in many of the tables available in the publications of National Statistics Offices. However, she misunderstood the law as an iron rule of fate which demanded such a number to occur. Thus she imagined that if as the end of the quarter approached the numbers were falling short, then more accidents would have to be engineered. Presumably also, some brake would have to be put on their occurrence if the numbers were mounting up too rapidly in the course of any quarter. It is not unknown to find people thinking in the same way today. But this is to get things back to front. The near constancy of the numbers from quarter to quarter is a *consequence* of an underlying statistical *process*, not an externally imposed law. We shall see that if one wanted to influence the death rate, it is the independence between what happens on successive days and the probability of the happenings that would have to be altered. If anywhere, God's involvement must be at the more fundamental level. This example makes it clear that the characterisation of a statistical law as a probability distribution may have been too hasty and I should re-trace my steps and dig a little deeper.

The quotation reproduced here is purely illustrative of her views. Much more could be gleaned from the article.

STATISTICAL LAWS: A DEEPER LOOK

I start with a very simple example which is far removed from the grand finale of the next chapter, which centres on the quantum uncertainties that some see as the prime arena for God's action. In a sense, I have already set off in the wrong direction by referring to statistical laws in chapter 3 as frequency (or probability) distributions but we can now begin to see that the matter is more subtle. There we met the idea of a frequency distribution, with a fixed shape, which described an aggregate property of something like time intervals between radioactive emissions. For a second pass I take what is, perhaps, the simplest example of all. The main point I wish to make is that frequency distributions which we may observe are not fundamental. What matters is the *process* by which the distribution is generated and it is there that the law-like character is to be found. This is the point which Florence Nightingale failed to grasp and which is clarified by the following example.

Many contests in sport, and elsewhere, depend on the toss of a coin at the start to determine which team or competitor has the initial advantage of going first, choosing end or whatever. In the long run this is seen as fair because, on average, the coin will fall one way half the time and the other way the other half. In individual cases it may give an unfair advantage to one party but in the long run such things will 'average out' as we say. Just occasionally one side may get the advantage on several successive occasions but everyone accepts that this is just 'the luck of the draw' and that no one can be held responsible for it.

There are actually several probability distributions associated with coin tossing. To begin with, and most obviously, there is that which is composed of the probabilities of heads and tails respectively. With a fair coin these probabilities are equal so we can picture the distribution as consisting of two 'lumps' each of size 0.5 located at 'head' and 'tail' as illustrated below.

The usual way of explaining God's involvement in such a sequence is to say that God chooses the probability, 0.5 in this case, but leaves open the way in which that proportion is achieved. This is where space is supposedly created for particular actions by God without disturbing the overall distribution. If we were to collect the results of a large number of tosses we would expect to get proportions close to 0.5 but the order in which they occurred could be almost anything. The problem with this statement is that it is not a complete specification of the statistical law involved. There are sequences of heads and tails which yield a frequency distribution of the right form but which are clearly not random; for example, one in which heads and tails alternate. This is not the sort of thing we see happening in tossing at the start of competitive games, so something is clearly lacking in this specification. It is that in coin tossing any toss is (has to be) totally independent of all other tosses. A higher being attempting to conform to such a law has more to do than merely ensure a roughly equal number of heads and tails; they have to be randomly mixed up as well.

A second law associated with the coin tossing process is obtained by looking at runs. To determine the length of a run

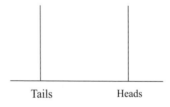

Tails   Heads

Fig. 8.1   The frequency distribution of the two outcomes in a single toss of a fair coin

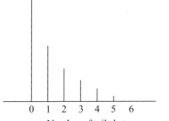

Number of tails between successive heads

Fig. 8.2 The frequency distribution of run length in successive tosses of a fair coin

of tails we count how many tails occur between successive heads. In the following sequence:

HTTHTTTHHTTTTHHH ...

The first run is two because there are two Ts after the first head and before the second head; the next run is three and this is followed by zero and zero again. The frequency distribution of runs will look something like that given above with frequencies declining in geometric progression. This is another statistical law associated with the coin-tossing process.

These two examples of probability distributions arising from the sequence do not exhaust the possibilities. Another is obtained by dividing up the sequence into blocks of, say, four outcomes each. In that case we could write our sequence as:

HTTH TTTH HTTT THHH ...

We may then count the number of heads in successive blocks which gives us the sequence: 2,1,1,3, ... The frequency distribution of these numbers, known as the binomial distribution, is a further statistical law associated with the same process. Such a distribution is illustrated in the figure below.

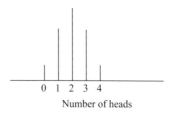

Fig. 8.3 The binomial distribution of the number of heads in four tosses of a
fair coin

It should now be clear that these three laws are derivative
in that they are all *consequences* of the basic probability law by
which the series was generated, namely that successive tosses
are independent and each has an equal probability of producing a head. If we wish to speculate about how God might be
involved in all of this we have to look at both the probability
and the independence, not at individual distributions.

KEEPING AND BREAKING STATISTICAL LAWS

We may get more insight into what is involved in keeping a
statistical law and of breaking it by attempting to put ourselves
in the position of a deity trying to generate happenings according to such a law. There are several possibilities depending on
what powers and objectives you have.

Imagine first that you are a lesser deity with responsibility
for overseeing the outcomes of sequences of coin tosses. Your
problem is how to decide the outcome of each toss. It is not
enough, as we have just seen, to ensure that heads and tails
occur in equal numbers. You also have to ensure that tosses are
independent. That means, for example, that any occurrence
must not depend on anything that has gone before in the
sequence. For example, the chance of a head must still be 0.5
whether it is preceded by one or a hundred heads – though this

may seem counterintuitive, that is exactly what independence implies.

There is one very obvious way to meet this requirement – by tossing coins or using some equivalent generating method. Later we shall see how this can be done on computers using pseudo-random numbers. This method ensures that the statistical law will be kept but it poses acute problems for the deity whose job it is to see that other outcomes, consequential on the toss, are in line with the overall plans of 'headquarters'.

But maybe such sequences are not generated by something equivalent to coin tossing. It is certainly not how those who emphasise the sovereignty of God would want to see it. For them the deity in charge would have to make a decision about every single outcome and, presumably, this would be done in the light of its consequences for all other happenings in the world to which it was connected. This might seem the easy option as it gives you complete freedom in choosing the outcomes. All that you have to ensure is that the aggregate properties of what you have done conform to the requirements of the random series. But this freedom is illusory and is not easy to achieve as I now explain.

THE DETERMINISTIC OPTION

The view that every single outcome is determined by the deity poses very serious problems because it implies that God's determinate actions successfully mimic chance and *at the same time* achieve other desired objectives. To give an idea of just how difficult this might be in practice, let us think through the process of trying to construct a sequence of heads and tails which had all the appearance of being random. Imagine the deity considering the options in advance with the aid of a set of black and white counters strung out in a line (black representing heads and white, tails, say).

To begin with there needs to be roughly the same number of black and white counters – out of several hundreds or thousands, perhaps. They must not be laid out in any recognisable pattern of course, because, if they were, the sequence would certainly not be random. However, we know that having the same number of black and white counters in an order which appears random is not the whole story. Randomness implies that what happens on any outcome has no effect on other outcomes. In particular this means that every black counter should be followed, equally often, by black or white. Similarly, every time we find two blacks in succession there should be an equal chance that the next outcome will be black. (There is a marked tendency for people to think that the longer a run of blacks, for example, the greater the probability that the next counter will be white. Sometimes it is said that the 'law of averages' requires it. It requires nothing of the kind.) So far I have considered two blacks in succession but it is already clear that it is quite hard to construct a series which has the characteristics of a truly random series – the reader who doubts this is invited to try!

The real problem emerges when we recall that the object is not merely to produce the appearance of randomness, which is difficult enough, but that each occurrence also has to trigger intended happenings in the complex web of the world. Imagine again that the deity surveys the sequence of black and white counters he has laid out and then considers what events they will trigger in the world at large. The second toss, maybe, leads to some undesirable consequence, so he changes its colour and finds, perhaps, that the consequential outcomes are more acceptable. But after he has done this a few times it begins to appear that the original randomness has been compromised, so he goes back to the beginning and starts juggling the members of the sequence to get a better overall fit. It is far from clear whether this exercise of the deity's considerable powers of foresight will converge on an acceptable solution. To make matters worse he realises that each choice creates a whole web of consequences, some of which are desirable and intended and others not. To show just how complicated things can become, remember that this exercise began by contemplating a series of coin tosses to decide, let us say, the choice of end in a football match. All sorts of things depend on the outcomes of football matches, outcomes which may be influenced by the winning of the toss. If they did not we would not be having this discussion. For example, a disgruntled spectator might leave the ground early in one case and be killed by a passing car whereas in the other he would arrive home safely. These two very different outcomes depend on the toss of a coin because which team kicks off leads to two different games. The death of that individual in the first game will have consequences for his family. A child who would otherwise have been born will not now be born and all the things that child might have done will not now be done. A neighbour might be so affected by the death that they turn to drink, or religion, with tremendous consequences for *their* families, including such trivial things as their decision to buy new shoes and the consequences *that* might have for the commission earned by a shop assistant . . . and so on. The ramifications are beyond our comprehension in number and scope. The ripples from that simple toss extend across the world and, maybe, to future generations. All of these have to be taken into account by the deity, along with all the consequences of the thousands of other tosses which fall within his responsibility. Steven Strogatz (2003, p. 189) reflects in a similar manner about the potentially profound consequences of the order in which we do up our shirt buttons. He also notes that the imaginative possibilities have not been lost on film makers, quoting the case of the film *Sliding Doors*,[3] in which there are radically different consequences for a woman's life depending on whether she gets through the sliding doors of a departing underground train before they close.

The response that only God is big enough to encompass all this and to choose the outcomes is too glib. The profound theological question is not so much whether God *could* handle the enormous complexity of the scenarios I have hinted at but whether it is a God-like enough thing for him to be doing. We return to the issue later and I do no more now than observe that the deterministic option I have been considering poses enormous complications which are often overlooked by those who so easily adopt it.

The whole situation can, of course, be looked at the other way round. Perhaps the sequence of heads and tails would have to be a *consequence* of decisions taken on other grounds. Thus, suppose that the decision of head or tail was based solely on the desirability of the consequences which that choice would initiate and that it just happened to turn out that the resulting sequence was indistinguishable from random. What we would have then would be something very close to the generation of pseudo-random numbers which we have already met in chapter 4. This possibility is certainly plausible but it does not avoid the problem of the immense complexity of trains of events triggered as the ripples spread outwards through the

---

[3] This is not the only example of the use of randomness as a literary device. The use can be more explicit as when the roll of the die is actually supposed to determine future outcomes. Luke Rhinehart wrote his novel *The Dice Man* in the 1970s and has followed it up with other works in the same vein, including plays and television adaptations. The English opera singer Lesley Garrett claimed that reading *The Dice Man* changed her life (*The Independent*, 18 November 2006).

web of possibilities. But even if this were the case, the statistical law in question would be the indirect *consequence* of all these other decisions and not the primary law.

It is worth pausing to reflect for a moment on the situation into which we shall have argued ourselves if we adopt this position. In effect we are saying that the outcome of the toss in a football match may be determined by the effects, long term and short term, which that insignificant happening will have on the future of the world. Is that really credible? It has to be, if we follow the logic of supposing the world to be determined, in every detail, by an all-powerful and all-knowing deity. Before committing ourselves to such seeming absurdities, perhaps we should examine another option.

OCCASIONAL INTERVENTION

This is, in fact, the third strategy to consider. The first, remember was to leave it all to chance, which seemed to leave little opportunity for God to act at all. The second, the deterministic option, seemed almost to strangle God in the complexity of his own creation. The third seeks to obtain the best of both worlds by combining a modest amount of direct action with a large dose of genuine randomness. In this third scenario there are many genuinely random happenings and, most of the time, the deity in charge has no more to do than to sit back and delight in the variety of the creation. On the rare occasions when intervention is necessary there will be nothing to give away the fact that intervention has occurred. Any actions are obscured by chance. This is because there is always some variation in the aggregate results of random processes such as coin tossing. Although about 50 per cent of all tosses are heads, the actual number will vary. It is this variation which conceals any intervention. Since the law of coin tossing does not specify *exactly* how many heads will occur, there is no way of detecting the intervention *provided that* that action is a rare occurrence. One might note, cynically perhaps, that this strategy escapes all criticism on strictly scientific grounds because it is empirically undetectable. The occasions of intervention are sufficiently rare for them not to disturb the regularity required by the statistical laws governing them.

However, this strategy is not without its problems, over and above that of wondering whether acting somewhat furtively under cover of randomness is a God-like thing to do. It still has to be possible for God to determine the points at which to intervene in the light of the consequences. These, as the illustrations used above showed, are extremely complicated and may not be universally beneficial. The real point at issue is whether it is actually *possible* to achieve sufficient control by intermittent intervention. Again, the glib answer that an all-seeing, all-knowing God can take such things in his stride, whereas mere mortals see only impossibilities, must reckon with the fact that even God cannot do what is logically impossible. And in this case we have no idea what can be achieved by occasional interventions in vastly complicated systems. Nevertheless we must allow this third strategy as a possibility, since it cannot, as yet, be categorically ruled out.

None of these three strategies offers a clear view of how God might act in the world in the space created by statistical laws. His total involvement in a deterministic world offers theological attractions to some but it is far from clear whether it would work. The same objections apply if the coin tosses are made purely in the light of their consequences. Occasional intervention is a possibility, but again, it is not clear whether or not it would work. Leaving it all to chance, as in the first strategy, deliberately rules out action in individual cases and so offers no help. I tentatively conclude, therefore, that the original hope that statistical laws would provide an opening for God's action was ill-founded.

RADIOACTIVE EMISSIONS REVISITED

The reader may feel that a great deal has been built on this single coin-tossing example, so, to reinforce the main message, we return to the case of radioactive emissions striking a Geiger counter. This will underline the importance of the primary process rather than any frequency distributions derived from it. Remember that the process involves the occasional emission of particles from a radioactive source. A record of such a process could be made in several ways but one of the most obvious is to record the times at which emissions occur. There are several probability distributions associated with such processes, each of which might be described, inaccurately as we now know, as a statistical law.

Another way is to construct the frequency distribution of time intervals between events. As noted above, this follows what is known as an exponential distribution, whose shape was illustrated in figure 3.1 (see p. 35). This exponential law does not fully determine the process as we can see if we try to use it to reconstruct the process. How would we do it and how might God do it? One way which might occur to us is to take a set of intervals whose distribution has the exponential form. We could place the intervals end to end and we would then have a series which conforms to the statistical law in the sense that it has the right frequency distribution. But a moment's thought shows that the distribution does not capture all that is implied by having a random series. Suppose, for example, that we began the series by starting with the shortest interval and then followed it with the next shortest, and so on. Clearly the result would not be a random series. In a truly random series the lengths would be all mixed up. This is because it is an essential feature of the random process that its future development does not depend upon its history. Given the proposed series we would be able to predict, at any stage, that the next interval in the series would be larger than all its predecessors. No such prediction should be possible. We might, therefore, go back and arrange the intervals in a random order to capture that particular aspect of the basic idea, which would certainly give us something much nearer to the truly random series. However, there would be more subtle differences even after both of these steps had been taken. For example, no finite set of frequencies would conform exactly to any distributional form. If the correspondence were too close, suspicions would be raised about whether it was too random. But this seems a complicated way of producing something which was supposed to be without any law or pattern. The similarities with the coin-tossing example are clear, even though one process is taking place continuously in time and the other discretely. The exponential distribution reveals one aspect of the process. Other aspects are revealed, for example by the Poisson distribution as we saw in chapter 3.

LAWS OF GENERATION

I have concluded that the lawfulness which is exhibited in what are widely, but somewhat misleadingly, called statistical laws is actually a manifestation of the underlying processes which generate the distributions. In the two examples considered above, these processes are best described by the constancy of a probability, or a rate, and by independence. These two aspects speak of a 'hands-off' approach to control rather than the expression of purposeful activity. If that is so, any purpose which they express must be via the *aggregate properties* of the system. In that case one only has to ask how a deity might generate something which bears all the signs of being left to its own devices. By far the easiest way to construct a truly random process is to use some kind of pseudo-random generator which is available on most computers. Although this would not be strictly random, it would be as close as makes no difference. In an important sense this is the simplest possible method. It does involve a constant striving to come as close to some ideal as possible. It follows that if there is a method so simple and elegant, it is surely the method God would use. There is nothing to be gained by imagining that even God, by taking thought about the timing of the occurrence of each event, could do any better. If there are to be random processes in the world to serve useful purposes, the creation of them needs to be in God's repertoire. There is then no need for him to be concerned with happenings at the individual level. All that matters is their behaviour in the aggregate.

MORE ON LAWS

Theologians often speak as though God decided upon the laws of nature, meaning that he decreed, for example, that the attraction between two bodies should vary inversely as the square of the distance between them. Each body in the universe is thus thought of as being endowed with properties which ensure that it obeys the prescribed law for ever. It is this lawfulness which is often claimed as a sign of God's involvement.

But, as was observed in chapter 3, there is another way in which people think about 'laws' according to which God is thought of as enacting laws, which have to be obeyed. It is worth noting, in passing, that miracles present no insuperable problem if we think of laws in this way. If God can make laws in the first place he can surely alter them, or even suspend them, for his own purposes. It is not at all obvious that this is the way to think about laws of nature. It certainly does not sit well with our focus on the basic underlying processes which generate the so-called statistical laws. We know that these laws emerge at the aggregate level out of lawlessness at the lower level.

To pursue the matter a little further, suppose that 'in the beginning' there was total chaos (a formless void as the book of Genesis puts it) at the micro level. Is it then conceivable that all laws, which we experience at a much higher level, are basically statistical? A positive answer could be viewed theistically or atheistically. The atheist might say 'I told you so. It needs absolutely nothing to create a lawful universe. It just could not be any other way. There is always order in chaos if only we look at the right level. It is not necessary to invoke a God as everything we observe could happen without one.'

The theist might counter by saying 'what could be more elegant than to conceive and get going such a simple and beautiful system? A universe which makes itself. A universe burgeoning with potential. Surely God could have done no other.'

It certainly does not seem to me that all laws have this statistical character, but there may be an inevitability about them which we have failed to detect. There may be something necessary about them so that not even God could have decreed them otherwise. Perhaps the laws with which we are so impressed are inevitable consequences of the very nature of things and that to expect them to be otherwise would involve a logical contradiction somewhere.

The manner in which order can arise out of chaos is a many-sided thing and it has already been explored in some detail in chapter 3. In chapter 4 it was noted that it is also true that most chaos, in the technical sense, is generated deterministically; that is according to some rule or formula. It would be unwise to conclude prematurely that, at bottom, the world is either random or deterministic. What we *do* know is that we are in a situation in which random behaviour at the micro level produces order at the macro level and where determinism at the micro level generates apparent randomness at the macro level.

In this chapter I have only scratched the surface of how God might control, or influence, things through the flexibility which chance offers. The world is full of probability distributions and I have considered the possibilities offered by only one. Quantum theory sees things on the grand scale and it is on this stage that most participants in the science–theology debate wish to play their parts. In the following chapter we shall see how far the ideas developed here carry over to that wider arena.

CHAPTER 9

God's action in the quantum world

Quantum theory provides the most fundamental account of the physical world and it is at that level, one might suppose, that the roots of God's action would be found. The fact that the theory is probabilistic leads to the hope that the uncertainties present will provide enough room for God to manoeuvre. The conclusion of this chapter is that this hope is too optimistic for very much the same reasons as those used in the simpler cases treated in chapter 8. It is argued that the transition from the superimposed states of the theory to the single observed states is most naturally explained by expressing the problem in terms of conditional probabilities.[1]

A STATISTICAL APPROACH

How God acts in the world, I repeat, has become one of the most discussed questions in the science and religion field – and beyond. As yet, there is no satisfactory answer to that question. This conclusion has been illustrated in the last chapter, on the small scale and on a restricted front. Now we view the

---

[1] My disclaimer in the preface applies with particular force in this chapter. Quantum theory is a highly technical subject and an area where a 'little learning' may be a particularly dangerous thing. Nevertheless if God does act at that level, it is important that non-specialists try to understand what is going on. Some theologians, notably Keith Ward and Nancey Murphy, have attempted to engage with quantum physics, and this contribution is offered in the same spirit. Quantum theory is probabilistic and that provides the point of entry for a statistician.

problem on a larger canvas to see how the conclusions reached there survive when they are translated to global, or even, cosmic dimensions. If we find that the conclusions of the last chapter carry over, there is little more to be said. However, the transition from the simple and idealised world of the last chapter is not automatic, because it is not clear whether the quantum world is probabilistic in the way that I have been supposing hitherto.

As noted in the last chapter, the most sustained treatment of the subject at a general level has been made by the Divine Action Project (DAP), which enabled many of the most distinguished workers in the field to collaborate over the period 1988–2003. In contrast to this collective contribution and, in some respects, in opposition to it, there is the major contribution of Nicholas Saunders in *Divine Action and Modern Science* (Saunders 2002).

This present contribution cannot be easily located within the framework established by the DAP and its aims are very much more limited. Our course has been set by our approach through the treatment of statistical laws in the last chapter, where I emphasised the importance of the underlying process which generates the laws. We approach the subject using the distinctive (possibly distorting) lens provided by the statistical/probabilistic way of viewing the world.

QUANTUM THEORY

The most fundamental account of nature is provided by quantum theory, which deals with the behaviour of matter at the level of electrons and other basic particles. The theory involves uncertainties and these seem to offer a way of seeing how God might act, out of our sight so to speak, to control what happens in the world. It is the fact that there appear to be irreducible uncertainties at the heart of nature which gives the subject its place in this book. If nature were deterministic to its very roots, it would be very difficult to find a place for what Saunders (2002) and others call *special divine action* (SDA). Particular acts of God, such as miracles and specific answers to prayer, would then require something quite drastic, for example the suspension or alteration of laws of nature. In such a deterministic world everything is pre-programmed and things work themselves out according to the inexorable logic of the machine. But in a world of uncertainties there may be some room for manoeuvre. Perhaps there is more hope of finding room for God to act if we allow for the probabilistic character of nature as described by quantum theory.

I must begin by saying something about quantum theory itself and why it might seem to be the arena, in part at least, for God's action. Classical mechanics is concerned with how things work on the human scale – with projectiles, cars and billiard balls, for example. It has been very successful in underpinning modern engineering and owes a great debt to Isaac Newton, with whose name it is often linked. However, if we move to things which are very small it is found that the world at that level does not conform to the prescriptions of the classical methods. Attempts to describe electrons and other entities at that level by classical methods do not work. This means that what happens at that level does not fit in with our intuition, which is based on the macro world in which our understanding is formed.

Quantum mechanics provides the mathematical machinery to handle what goes on at the atomic level. It was one of the great intellectual achievements of the twentieth century and has proved remarkably successful in dealing with the world as it is when observed at what I shall loosely call the micro level. Readers who wish to learn more might well begin with John Polkinghorne's *The Quantum World* (Polkinghorne 1984) or with the introductory articles available on the World Wide Web. For example, Wikipedia, the online encyclopaedia, has several relevant articles at http://en.wikipedia.org/wiki/ Quantummechanics, which links to other similar articles including one on the 'Interpretation of quantum mechanics'.

The interest of quantum mechanics, for present purposes, lies in the fact that it is a probabilistic theory and it is this fact which offers a toehold for the theologian. If certain happenings are unpredictable they may provide scope for God's action. Some theologians, in fact, think that all of those (potential) happenings must be within God's immediate control.

It is necessary to emphasise that quantum theory is deterministic in one important sense. A quantum system is represented in the theory by something called the *wave function*. In relation to a particle this enables the probability of its being found at any particular location to be determined. In other words, it does not tell us exactly where the particle is, but how likely it is to be found at any location. Its position is, essentially, described by a probability distribution. A basic equation of quantum theory tells us how this probability distribution changes over time. There is nothing uncertain about this. It is a deterministic differential equation not so very dissimilar to those differential equations which chart the development of deterministic systems at the macro level. If all we want to do is map the future of a system in terms of probability distributions of the quantities of interest, chance does not come into the picture.

Quantum systems include things which are called 'observables', which, as their name indicates, can be observed, indirectly at least. This means that they can be made to trigger an observable event such as the clicking of a Geiger counter. What we observe is not the probability distribution but a value sampled at random from the probability distribution. That is a statistician's way of putting it. In the language of quantum physics it is spoken of as the collapse of the wave function or wave packet. (The precise meaning of this is important and we shall come back to it later.) What we actually observe is something which tells us, for example, whether the spin of an electron is 'up' or 'down'. The underlying theory will only tell us that it is up with a certain probability and down with the complementary probability. This is described by physicists as the 'superposition' of the two states; in statistical parlance it is a mixture. The act of observation selects one of these options with the specified probability.

It is only at this final stage that chance enters. The selection appears to be entirely at random. That is, there is nothing that we could possibly observe which would help us to predict the outcome. If God acts at all at this level, the argument goes, he must determine, or at least influence, these outcomes.

GOD'S ACTION AT THE QUANTUM LEVEL

Many theologians and sympathetic scientists have seen the attractiveness of locating God's action in the world at the quantum level. This may be due partly, at least, to the difficulty of seeing how God could act on any larger scale in the physical world. But more important, perhaps, is that a comprehensive theology requires an account in which God is not excluded from any part of his creation. Since what happens at this level is not entirely predictable it is natural to fill the causal gap with the action of God. Although this is an attractive proposition, its explication is fraught with difficulty, as we shall see shortly.

The general problems of assuming that God acts at the quantum level have been clearly documented by Saunders

(2002)[2] among others. Brecha (2002), in particular, writing as a practising scientist, thinks that the room for God's action at the quantum level is less than some theologians believe. The tendency of some to treat quantum theory as a metaphor fails to come to grips with the realities of the situation. The crux of the problem, as I hinted above, is that there are several competing interpretations of quantum theory. Polkinghorne (1984) discusses four interpretations and Saunders seven. These interpretations are all consistent with the scientific evidence, so there is no means by which we can know how things 'really are'. And if we do not know that, it is hard to see how a single theological account can cover all possibilities. The issue is sometimes prejudged by speaking of 'quantum events' as though they were realisations of laws expressed by probability distributions. It is not even known whether the reality which quantum theory describes is, in fact, probabilistic; some accounts are deterministic and thus offer no foothold for theistic involvement. Among those on the deterministic side, Einstein was very reluctant to accept that chance is a real element of the world at this level. Others, such as David Bohm,[3] have also sought to find deterministic explanations of the formalism. These often seem rather contrived but, if they are correct, there would be no place for God's action under the guise of chance.

[2] In addition to his book, Saunders wrote an article entitled 'Does God cheat at dice: divine action and quantum possibilities' (2000). This anticipated the publication of the book referred to in the text. It was followed, in the same journal, by the paper by Brecha (2002) and a direct response from Ward (2000) taking issue with Saunders' interpretation of what he had said in his book *Divine Action* (Ward 1990).

[3] David Bohm (1917–92) was a distinguished, if unusual, physicist who challenged the conventional understanding of quantum theory. He proposed a non-local deterministic theory to account for quantum behaviour. Hodgson (2005) also favours interpretations of a deterministic kind.

Two questions now arise, one theological and the other scientific. Is it reasonable to suppose that God could act at the quantum level? And secondly, if he could, would this be an effective way of influencing what happens in the macro world? The two questions are closely bound up together and the second is critical because it is hard to see why God would want to use an approach which was not effective. Roughly speaking, happenings at the quantum level can make themselves felt at the macro level in one of two ways. First, and predominantly, they contribute to the average effects of the large ensembles of particles. Secondly, a micro event might trigger a sequence of causally linked events which produces something detectable at the macro level. Furthermore, we know of one way, at least, in which this can be done. Chaos theory shows how the outcome of a deterministic, non-linear process may depend critically on the initial conditions. In this way, a micro happening may be sufficient to produce macroscopic changes. In principle, therefore, it might be possible to 'steer' things in the way the Designer wished.[4] It is, however, far from clear whether this would actually work in practice, given the enormous complexity of the world.

One of the most puzzling things about quantum theory is the seemingly paradoxical effect of making an observation. Before observation, the situation is described by what physicists call the superposition of several events. After the observation is made, one of those events is realised. How does this come about? The story of Schrödinger's[5] cat was constructed

---

[4] It is too readily assumed that any gap which is revealed in the causal nexus at the quantum level could be exploited by the Deity to achieve any desired outcome in the world. I am not aware of any attempt to map out the possibilities.

[5] Erwin Schrödinger (1887–1961) shared the Nobel prize for Physics with Paul Dirac in 1933. He succeeded Max Planck in Berlin in 1927. He left Germany

to highlight the paradox. Looking at it from a statistical perspective, it does not seem quite so paradoxical after all.

A STATISTICIAN'S VIEW OF SCHRÖDINGER'S CAT

Quantum mechanics, as I have noted, provides an excellent means of calculating the behaviour of systems at the level of elementary particles. The problem of interpretation is concerned with what reality is actually like at that level. As already noted, it is impossible to picture it in terms of the macro world with which classical mechanics deals. This does not matter if all we wish to do is make calculations, but if we wish to see how God might act at the micro level, it becomes of central importance. Quantum theory describes the micro world in terms of what are called superpositions, that is, mixtures of states, but when an observation is made only one component is actually observed. How the process of observation reduces the mixture to a single component has been something of a conundrum. The story of Schrödinger's cat was invented to help to elucidate the problem.

Probability is central to quantum theory and this provides the point of entry for a statistician. Statisticians regularly handle situations where the formalism is much the same, and it is interesting to see how the switch to another language alters the perspective if not the substance.

The state of affairs in the micro world is given by the wave function. As Max Born showed, this can be interpreted in probabilistic terms by telling us, for example, how likely a particle is to be found at different locations. In contrast

on Hitler's rise to power and subsequently held various appointments until taking up his final post at the Institute for Advanced Studies in Dublin, from which he retired in 1955.

to classical mechanics, we can thus only make probabilistic statements about location. Shrödinger's equation describes how this wave function changes over time. The way that the probability distributions change over time is thus entirely predictable. Thus far in the argument, there is no unpredictability in which God's action might be cloaked.

Chance enters at the point where observations are made. I have already noted that it is possible to observe things at the macro level which indicate what is going on at the micro level. As an example, I took the case of an electron passing through a magnetic field so that its path will be diverted in one of two directions according to whether its spin is up or down and the new path of the electron can be detected by placing a Geiger counter so as to intercept the particle. We do not actually observe the spin itself but some consequences of it which are detectable at the macro level. It is here that the nub of the matter lies. All that quantum theory tells us is that the spin is equally likely to be up or down. The state of the electron is then said to be a superposition of the two states – up and down but we can only observe one of them. But how and where does this happen? This is where Schrödinger's cat comes into the picture – not to solve the problem but to highlight what it is. In the quantum world superpositions (mixtures) can exist; in the world of observables they cannot. In some mysterious way, it seems, the act of measurement turns the mixture into one of its components. Let me restate the problem, as Schrödinger did, in terms of the cat.

One can even set up quite ridiculous cases. A cat is penned up in a steel chamber, along with the following diabolical device (which must be secured against direct interference by the cat): in a Geiger counter there is a tiny bit of radioactive substance, so small that perhaps in the course of one hour one of the atoms decays, but also, with equal probability, perhaps none; if it happens, the counter tube discharges and through a relay releases a hammer which shatters a small flask of hydrocyanic acid. If one has left this entire system to itself for an hour, one would say that the cat still lives if meanwhile no atom has decayed. The first atomic decay would have poisoned it. The psi function for the entire system would express this by having in it the living and the dead cat (pardon the expression) mixed or smeared out in equal parts. It is typical of these cases that an indeterminacy originally restricted to the atomic domain becomes transformed into macroscopic indeterminacy, which can then be resolved by direct observation. That prevents us from so naively accepting as valid a 'blurred model' for representing reality. In itself it would not embody anything unclear or contradictory. There is a difference between a shaky or out-of-focus photograph and a snapshot of clouds and fog banks. (Schrödinger 1935, trans. Trimmer)

Immediately before the box is opened we do not know whether the cat is alive or dead. It is equally likely to be in either state. In quantum terms its state is a superposition of the two states alive and dead but when the box is opened, it is only possible to observe one of the constituents. How does the act of observation resolve the mixture into one of its components? The wave-function description is, in this case, a mixture of two states with probability one half associated with alive and the other half with dead. After observation it is reduced to a single point – all of the probability is loaded onto one state. This is referred to as the 'collapse of the wave function' or the wave packet. The wave function has been changed in form, it appears, by the act of observation. How can this be, and does God have any hand in it?

At this point I shall retrace my steps and recast what has been said in terms more familiar to a probabilist or statistician as set out in chapter 5. The crucial point to remind ourselves of is that, in technical terms, a probability is a function of two arguments. In non-technical terms this means that the probability in question depends on something that we already know. An example will illustrate what is meant. The probability of throwing a six with a fair die is 1/6. But if you are given a doctored die, on which the one has been changed into a six, the probability will be raised to 1/3. If you tossed this die without bothering to inspect it you would, justifiably, say that the chance of a six is 1/6. But if, after inspecting it, you see that there are two faces showing six, this new information would lead you to revise your probability. The probability thus depends upon what you already know. Again, a life table will give the probability that a fifty-year-old man will be dead before reaching the age of sixty, and insurance companies base their policies on such information. However, if a medical examination reveals that the man has cancer of the liver, the probability will be changed because of that extra information.

Ideally the notation we use for a probability should include a statement of what is assumed to be known. In practice this is often omitted because it is assumed to be understood. Things which are 'understood' are easily overlooked, but here we must make them explicit. The probability of the up and down alternatives is understood to be a half in each case. This can be empirically verified by making a large number of observations and verifying that the two alternatives occur equally often. This equality is something which applies to electrons in general and must be presumed to apply to those we are about to observe, in particular. When we pass the electron through a magnetic field, we acquire some new information – in which direction it is diverted. The probability that its spin is up or down is changed by having this extra information. It becomes one for up, if up is what we observe. The collapse of the wave packet is thus marked by the transition in the probability resulting from the acquisition of extra information.

In general, the acquisition of any relevant information changes the probability distribution. But how and when does this happen? This is the problem the cat example is designed to clarify.

So to answer these questions let us return to Schrödinger's cat. Just prior to the box being opened our *state of knowledge* about the cat is the mixed probability distribution which assigns equal probabilities to each state. When the box is opened we discover whether or not the poison was released. If it was, the cat will be dead. Our initial two-point distribution (dead or alive) collapses onto one of them. The relevant probabilities now become those conditional on our knowing what has happened while the box was shut. In general a total or partial collapse of the wave function is caused by the acquisition of new information. Where, then, is the wave function and where does its collapse take place? The answer is – in our heads. The wave function is a mental construct embodying what we know about the system. This explanation is not new of course; in fact it was the explanation that Max Born favoured and it follows simply and obviously from interpreting what is going on in probabilistic terms. It is also the first of the four possible interpretations which Polkinghorne gives in *The Quantum World* (1984, p. 63). However, he promptly dismisses it (noting in passing Max Born's adherence to it) on the grounds that it reduces the august and objective science of physics to a branch of psychology. Presumably this is because a mental construct in an individual's head does not seem to have the hard reality of a feature of the real world. This view is based upon a misunderstanding of what is intended. The further knowledge provided by the viewing of the contents of the box is not a private subjective thing; it is publicly available. It is part of what is given to anyone who chooses to make the observation. The idea is not unfamiliar in other fields of application. For example, calculation shows that the conditional probability that a person immunised against influenza will succumb is smaller than the unconditional probability that someone not immunised will contract the disease. The incorporation of the knowledge that immunisation has taken place does not make epidemiology a branch of psychology. Conditioning *may*, of course, introduce a subjective element into scientific reporting. For example, if the results of the cat experiment are not directly observed but reported by an intermediary, there will be two bits of information to be incorporated into the collapsed probability calculation. First, there will be the factual report that the cat is alive, say. Secondly, there will be the intermediary's reputation for telling the truth. The latter part is a subjective matter which may mean that different reporting will lead the wave function to collapse on different values. But if the evidence of the cat's state is open to general observation, the subjective element becomes swamped by the common core of truth.

This leaves the question of when the actual collapse takes place. For any *individual* it must happen when the information is received and processed. There is, therefore, no unique time. It cannot, however, be earlier than the event which triggers the death of the cat. If there is a radioactive emission, then the time of its occurrence determines everything. This will be either when the emission takes place, or when it fails to happen (that is, at the end of the period). In the standard discussion of these things, the making of the observation is spoken of as the act of measurement. Since measurement takes place in the macro world, it can only take one of the two superimposed values. In the case of the cat, the opening of the box is the act of measurement which yields one of the observations alive or dead. A statistician would use different terminology – which does not change anything – but it does somewhat modify the strangeness of saying that the act of measurement brings about the collapse.

Prior to observation our state of knowledge is summarised in a probability distribution. Afterwards, we have a value from that distribution. It is as though, at that moment, we draw a sample at random. In this perspective the crucial act is therefore one of random sampling. In the jargon of probability theory the characteristics of an electron, say, before observation, are described by *random* variables.[6] Such a variable is not regarded as a single number but consists of a set of numbers, each having a probability associated with it. Drawing a sample is thus a matter of selecting one of those values with the associated probability.

What is going on may be looked at in a slightly different way, of which Einstein evidently approved. According to this view, what quantum theory describes is not a single process but an aggregate, or ensemble, of processes. It does not tell us which one is the case at the moment we choose to observe it. Because we can only observe one particular value of anything we set out to observe, rather than a mixture, our sampling mechanism is a method of making a random selection. The act of measurement is an act of sampling.

If this interpretation is correct, the best account we have of the micro world is no better than a probability representation.

---

[6] The notion of a random variable is fundamental in probability theory. In algebra, letters or other symbols are used to represent numbers. We are used to the idea of replacing symbols by numbers in a formula in order to obtain the numerical value of some quantity. For example, there is a formula which tells us what temperature on the Fahrenheit scale corresponds to a given temperature on the Celsius scale. Knowing the Celsius temperature enables us to calculate the corresponding Fahrenheit value. Probability theory deals with quantities which do not have a single value but may take different values, each with known probability. When a symbol representing a random variable appears, it does not represent a single value but a set of values. It may help to think of an associated probability mass 'smeared' over the set of values giving different weights to each.

Whatever might be involved in the selection of these particular observed values, this does not affect the description of the world at large, provided by the wave function. In statistical terms, therefore, nothing is changed. The macro world goes on just as before on the back of the micro-level processes. Making observations breaks through the probabilistic fog at one particular point but it does not change the world. A few observations by a few people on this particular planet do not change how the world was – and is. Measurement does, however, increase our knowledge of little bits of the world.

THEOLOGICAL ISSUES

Where does God come into all of this?[7] Does he observe everything and thus sample everything that is going on in the whole universe? If he did, he would simply have the total aggregate picture of the world, which would be the statistical version of what we describe in terms of random variables and probabilities. A simple example may make the matter clearer. If we measure the lives of a few dozen electric light bulbs we may notice that the frequency distribution of their lifetimes has a shape close to the normal, bell-shaped, curve which we met in chapter 3. An examination of the mechanisms of failure might confirm this view and lead us to determine a guarantee period on the basis of this assumption. If observation by God of all such potential experiments led him to an aggregate normal distribution, he would have, in aggregate statistical terms, what our random variable model described. God's complete knowledge of what is going on does not, essentially, alter the

---

[7] Those who wish to pursue the theological interpretations which may be put on quantum theory may find the article by Roger Paul and the response from Rodney Holder useful (*Science and Christian Belief* 17 (October 2005): 155–85).

reality of the universe. God's existence does not, therefore, make any difference to our physical account of the system.

Many physicists are not very concerned with these philosophical and theological issues but use the theory in a purely instrumental fashion. Theologians and theologically minded scientists, on the other hand, have, perhaps, been over-eager to see scope for God's acting in the uncertainties of the quantum world. One of the earliest was William Pollard whose book *Chance and Providence* (Pollard 1958) has been very influential. More recently Robert John Russell and Nancey Murphy have been prominent among those who have argued that God acts in the uncertainties of the quantum world – but not only there.

I have concluded that quantum theory describes the world in probabilistic terms. In principle, therefore, it is not essentially different from the simpler situation considered in the last chapter. There we were concerned with one probability distribution with a single observable outcome; here we have an immensely complicated probability structure involving incomprehensibly many variables. But in both cases the theological question is much the same: how can God be understood to act in such an environment to achieve particular purposes in the world? Once again there are the same three directions which an explanation might take:

(a) In reality there is no uncertainty. All is determined by God and what we see merely appears random to us. If we could comprehend God's thoughts we would see that everything that was going on served some determinate purpose.

(b) In reality, what appears random to us is exactly what it appears to be. A vestige of determinism might be preserved by allowing that all of this randomness might be generated deterministically as are pseudo-random numbers.

(c) While (b) might explain most of what happens, God has the ability to monitor what is going on and, very occasionally, to intervene to steer the cosmos in some desired direction.

Here, as in the last chapter, there are serious objections to each of these possibilities, which I take in turn.

On (a) I have noted that the triggering event at the micro level and the consequent happenings along the way will also act as triggers for countless other causal paths, each having consequences in the macro world. Some of these outcomes may be harmless but others might have effects which cancel out, or work against, the intended outcome. All of these possibilities would have to be contemplated and allowed for. Further, we are not thinking of one macro outcome but untold trillions over aeons of time. The bizarre picture of God seated in front of a celestial control panel watching microscopic happenings throughout the universe and reacting to them almost instantaneously may be logically possible but it hardly fits with the notion of the loving heavenly Father of orthodox Christian belief, neither does it accord with our idea of how high-level control should take place. The analogy of the management of a large company might be more helpful here. The chief executive is not concerned with the day-to-day details of office and shop-floor activity. They are left to subordinates with more limited responsibilities. The chief executive's job is to focus on the big questions of policy and the strategic issues. There is a simplicity and elegance about the management structure of the firm which is lacking in the picture of a celestial manager with a finger on every button. One would surely expect the God-directed activity of steering the universe to be no less impressive than those devised by mere mortals for their own organisational creations. The picture of a world in which the details take care of themselves, leaving the big issues to the Creator, is more appealing and more worthy of directing our worship. This, perhaps, is a case where we are too prone to see God in the image of man as someone who thinks control depends on overseeing every detail.

On (b) Stephen Hawking (quoted in Saunders 2002, p. 128) makes a very shrewd point with which I conclude the discussion of whether what appears to be random is really random. 'If one likes one could ascribe this randomness to God, but it would be a very strange kind of intervention: there is no evidence that it was directed towards any purpose. Indeed if it were, it would, by definition not be random.' It is part of the last sentence of the quotation that touches the nub of the matter. Randomness is what we have when all purpose and direction is excluded. We cannot, therefore, smuggle purpose in by the back door under cover of randomness.

If, on the other hand, the equations of quantum theory do describe genuine randomness, there is no room for action by God mediated through individual events at the quantum level. This leaves open the central question of this book which is whether other things are achieved by this very randomness which are equally expressive of God's intentions.

It is because (b) has rarely been seen as an option that the third of the earlier alternatives has proved so attractive. After all, perhaps all that is needed is for God to give a nudge from time to time, at critical junctures, to keep things on course. If these interventions are sufficiently rare there is no chance of them being detected, and it seems that we have the best of both worlds. The lawfulness of nature is preserved and space has been created for God to act. The objections to this supposition are essentially the same as those advanced in the last chapter for the much simpler situation considered there. First, it is far from clear that sufficient control can be exercised in this manner. If anything, the problem is compounded here by the much greater complexity in the world at the quantum level. Secondly, there is the theological question of the furtiveness of this manner of acting, which promptly raises the question of why God could not have got it right the first time. Finally, if the notion of special divine action (SDA) is to be located in the uncertainty of quantum processes, it is clear that this is neither easy nor likely to be very effective.

Another interesting theological question arises at the point of sampling, or measurement. The account I have given says nothing about how the sampling is done or how the random variable is turned into an observed numerical value. This is essentially the same problem as in chapter 3 with reference to radioactive emissions. It is also the problem which the story of Schrödinger's cat was designed to illustrate. In chapter 3 we saw that there was nothing observable which would help us to predict the outcome. Could it be God, therefore, who decides what we shall see? Is it God who determines, at each point of measurement, what value is actually observed? Maybe he does and maybe he does not, but it is worth asking what theological benefits are to be had by an answer to that question. If he does not determine what happens, then clearly he is not acting at such points. If he does, what purpose could be achieved thereby? We have already noted Hawking's remark that chance rules out purpose and vice versa. If chance enters only on those occasions where we make a measurement, our knowledge will be affected but the overall effect on the progress of events in the world will be negligible. It is true that the knowledge gained by us might influence our future actions and this could, conceivably, be how God interacts with us, but this seems a somewhat cumbersome way of going about things. For the present, the conclusion has to be that it is very doubtful whether there are any quantum events which God could influence whose outcomes might significantly determine what happens at the macro level.

Moving on from this point, we come again to the problem of saying how genuinely random events can be *caused* if they are truly random. Humans clearly can make choices and, as a result, affect what happens in the world. Can God do the same, and if so, how does he do it? As noted before, such events must either be caused, in some sense, by God or be without any cause at all. So the question really boils down to one of whether God acts from *within* by upholding and directing everything that happens or whether he somehow acts *on* the creation from outside by choosing from the myriad possibilities on offer. These two possible modes of action are often described as bottom-up and top-down. They are not mutually exclusive, of course, though how they might be brought together is not as straightforward as it might seem. Some clues might be obtained by looking more closely at how *we* use chance and this we shall do in the next chapter.[8]

---

[8] One of the most sustained and detailed discussions of God's action at the quantum level is given by Rüst (2005). There is no space here to give a full summary and critique but the following quotations will give an indication of the author's position – which differs fundamentally from that advanced in this book:

'God occasionally uses a selection of specific outcomes in quantum and other random events, in order to guide natural processes in the desired direction' (p. 197).

'Wherever there is randomness, science cannot distinguish between truly random events, providentially decreed by God, and specific events, selected by his creative choices' (p. 197).

'This suggests some divine guidance of quantum and other random events' (p. 198).

'God has plenty of options to providentially and creatively direct (or override if necessary) both natural events and actions of personal free-will creatures' (p. 200).

'God's "interference" may even represent a specific probability density function modifying the natural stochastic one' (p. 201).

CHAPTER 10

The human use of chance

Chance is no threat to the notion of design as is evident from the fact that we actually use it to achieve our own purposes. Some of these uses are long established but the ready availability of massive computing power has opened the way for a much wider range of applications. We are now able to generate randomness on a large enough scale to simulate many complex processes. Applications range from chemistry to music, with statistical sampling still playing a central role. Competitive situations call for chance selections in the choice of strategies, and so-called genetic algorithms attempt to mimic design by random variation and natural selection. These and other uses of chance are surveyed in this chapter.

IS CHANCE USEFUL TO US?

So far we have thought of chance as part of the natural order of things. Whether real or not it pervades the natural world, where many see its presence as a threat to the sovereignty of God. Eliminating chance as a possible explanation has been one of the main objectives of the Intelligent Design movement. Before I move on to argue that chance is a friend, not a foe, it will be helpful to prepare the ground by noticing that chance is often 'man-made' with the deliberate intention of achieving serious objectives or even entertaining us. This consideration may help to prepare us for the altogether more radical idea that in so doing we may be imitating God himself.

We begin with the more frivolous end, as some will see it – namely games of chance. Games are recreational activities designed to entertain as much as to educate and they come in several varieties. There are games of pure chance, where the outcome is determined solely by the role of a die or the dealing of cards and where skill plays no part at all. In other games the outcome may depend on a mixture of chance and skill, where the role of chance may be minor or major. In many team games such as cricket, for example, a main role of chance may be to decide, by the toss of a coin, which side has the choice of batting first. In card games, such as bridge, much more may depend on chance through the dealing of the cards.

Not all games are played against a single, or several, opponents in a competitive fashion. Playing, as it is rather oddly called, in a national lottery or on a fruit machine in an amusement arcade is a solitary activity which involves no skill at all – though there are many pundits who purvey advice as if it did.

Why is game playing such a widespread and, for those who organise it at least, lucrative activity? There are people who will not take part in games which involve chance, though few, if any, carry their scruples so far as to avoid them altogether. Chance introduces an element of uncertainty into games and this, in turn, induces surprise, excitement and enjoyment without the attendant risks and costs of real-life risk taking. For a time players live in an artificial world in which they enjoy the positive benefits of chance and derive immense satisfaction from it. For some, of course, this is not sufficient and they seek to inject the costs of real life into the action. The gambling instinct is deep rooted and the extra excitement generated by the real risk involved may, in moderation, add to the perceived recreational value. However, the line between the positive and the negative benefits is hard to draw and it would be foolish to pretend that man-made chance, in this context, is wholly beneficial. This should not be allowed to detract from the harmless pleasures which many derive from board games such as Monopoly, and Snakes and Ladders. The point is simply that man-made chance plays a useful role in society and adds greatly to the total human experience.

There is another contribution which chance makes in the world of games. In chapter 8 I considered the tossing of a coin at the start of a football match, to give a choice of ends. Here I shall delve a little more deeply into what that implies. It is not primarily to inject a further modicum of surprise into the match – though it may do that. Its prime purpose is to ensure fairness by eliminating any advantage which the local features of the pitch or environment might confer on one side. If the decision were made by any person, that individual could be accused of bias. If the choice depends on the toss of a coin there can be no argument. The principle at stake is one of *fairness* but it is worth pausing to ask in what sense it achieves this. For, in the end, one side has the advantage and this may well contribute materially to the outcome of that particular match. There are two answers which can be made to this query. One is by appealing to the 'long run'. Any particular side or individual will, in the long run, get the advantage on about half of the occasions. This could, perhaps, be achieved by more deterministic methods but only at the price of much greater organisational demands. Tossing a coin on each occasion achieves much the same end more cheaply and simply.

The second way of justifying the random choice of starting position is that it eliminates all possible human bias. There can then be no justified claim of human interference in the outcome. Furthermore it is important that the tossing should be public, with both sides witnessing the event. All of this can be ensured by the simple tossing of a coin.

There is another aspect to fairness. Games of pure chance eliminate any disparities there may be between the contestants. At first sight this might appear to run counter to the whole idea of competing but it is a way of bringing children or others of limited ability into a contest on equal terms. The benefits conferred by the shared excitement of game playing can thus be appropriated by a wider spectrum of individuals. Even when some skill is involved, the weaker partner still has some chance of winning as a result of 'the luck of the draw'. Chance may not level the playing field entirely but it does help to make the game more interesting for the less able players.

CHANCE CAN HELP TO SOLVE GEOMETRICAL PROBLEMS

Paradoxically, chance can be used to solve problems in mathematics which may have nothing to do with uncertainty. As an example, imagine that we wish to calculate the area of an irregular land mass such as, for example, the Peloponnisos in Greece or Newfoundland in Canada. One rather crude way is to trace the outline onto squared paper and to count the squares lying inside the outline. If the squares are small enough, this will be time consuming but will give a reasonably accurate answer. Another way is to superimpose the map on a square which is large enough to include the whole area. If we adopt the side of the square as our unit of length we can choose points at random within this unit square. (Exactly how we do this is an important practical matter but for present purposes it can be ignored: we would choose pairs of random numbers within the range (0, 1) and each pair would define a point within the square. Sticking in a pin with eyes blindfolded is the vernacular equivalent.) The proportion of the square occupied by the land in question can then be estimated by the proportion of randomly chosen points which fall within the land area. This is only an estimate, of course, and its precision is determined by how many points are selected – the more points, the more accurate the estimate. The precision can be estimated by using sampling theory. We can then calculate in advance how many points we shall need to select to achieve a desired precision. This can all be automated on a computer and would be extremely fast in practice. Here we are using chance as a tool to solve a mathematical problem: there is no uncertainty in the problem; it all lies in the method of solution. A great merit of this approach is that it is not confined to two-dimensional objects but will work in three dimensions, where the squared-paper method cannot be used. All that we need is a way of determining whether or not a point lies within the object.

In the above example we are using the fact that what is random at the level of the individual point creates regularity when the points are aggregated. The same principle applies in many other areas. A spray of paint, for example, consists of a mass of tiny droplets which are distributed more or less randomly throughout the area covered by the spray. A short burst of spray will produce a spotty effect with unequal amounts of paint being delivered to different areas. But a longer burst will produce a much more uniform coverage because the number of droplets per (small) unit area will now, proportionally speaking, be more nearly the same. In practice this method will be much better than what could be achieved by hand.

MONTE CARLO METHODS

This allusion to the Mecca of games of chance may raise unreal expectations.[1] We shall not be concerned here with methods

---

[1] The name Monte Carlo comes from the Principality, in which the casino is a centre of randomness. Stanislaw Ulam, a Polish born mathematician, is

of breaking the bank of that Principality but in using chance to imitate, or simulate, physical or social processes which are too complicated to study mathematically. Here we move from the realm of simple and rather contrived examples to processes which play a major role in almost all branches of science, pure and applied. To illustrate what is involved I shall use a very simple queuing example.

Imagine someone serving behind a counter, as in a shop or post office, with a line of customers before them in order of arrival. Uncertainty about what happens subsequently may arise in two ways: there may be variation in the times taken to serve each customer and there may be irregularity in the times at which new customers arrive. This is a dynamic process and the state of the system will change as time passes. We may want to know such things as the proportion of time the server is occupied and the average waiting time of a newly arrived customer. To answer such questions we may turn to mathematics[2] and, if the uncertainty can be described in simple enough terms, the theory of queues may enable us to do just that. Often this is not the case and then we can turn to Monte Carlo, or simulation methods. These provide us with a means of generating service and arrival times that have the same probabilistic patterns as those observed to occur in the real process. If we do this many times we shall begin

usually credited with inventing the technique in 1946 though, like so many such inventions, it is easy to trace the essentials of the idea long before that. Simulation is another term used in almost the same sense, although it does not necessarily carry the implication that the process involves a chance element.

[2] Queuing theory was much in vogue in the 1960s as one of the techniques of operational research. One of its main applications was to studies of traffic intensity in telephone systems, which goes back to the early part of the twentieth century. Most real queuing systems are much too complicated to be amenable to mathematical treatment, and resort has to be made to simulation.

to build up a picture of the distribution of waiting times and suchlike. Furthermore we can experiment with the system by, for example, looking at the effect of adding a new server. In effect we are using chance to imitate the process and so to learn something about the real world.[3]

This queuing situation is actually much simpler than most that occur in real life and its behaviour is fairly well understood. Monte Carlo methods really come into their own when the system is much more complicated. For example, congestion problems of some complexity are common at airports, on the roads and in many industrial production and distribution processes. However, it is the principle of the thing which concerns us here. Without the ability to simulate complex processes, and to do so rapidly enough to accumulate information while the problem is still a live one, ignorance and hunch would still be the only guide in many decisions.

Much of the work on networks touched on in chapter 8 has been facilitated by the use of Monte Carlo methods (see Barabási 2003). It is relatively easy to simulate the behaviour of a network on a computer. The results often give clues as to what results might be proved mathematically. It is usually much easier to prove something when you already have a good idea of what the answer is.

Essentially, it is the speed and capacity of modern computers which has revolutionised this field. For present purposes it clearly demonstrates how chance can be useful for learning about the real world and achieving things within it.

---

[3] A similar kind of example showing how natural selection can generate novelty is given by Ayala (2003), beginning at the bottom of page 19. This shows how bacterial cells resistant to streptomycin can be produced by using natural selection to sift through a vast number of possibilities.

RANDOM CHEMISTRY

This term seems to have been coined by Stuart Kauffman in the early 1990s though it is also referred to as combinatorial chemistry. It does not, as the name might suggest, refer to chemistry carried out in a haphazard fashion but to the systematic use of randomisation to discover complex molecules capable of combating disease. In the search for effective drugs one needs to construct molecules capable of binding to invaders to neutralise them. One might imagine doing this constructively by identifying what sort of shape the molecule needs to have and then building the required drug to order. This is more easily said than done. If the basic ingredients (chemicals) are known, the number of ways in which they can be linked together to form a possible drug may be truly enormous. This number may be so large that, even if it were possible, there would not be sufficient time before the end of the world to sort through them all systematically and find any that were effective. Some other strategy has to be found of searching the 'space', as it is called, of possibilities. This can be achieved quite easily if there is some way of mixing up the ingredients so as to produce a great many, if not all, of the possible candidates. In addition, we need to have some means of identifying any which 'work' – because these will be the potential drugs. The proposed method works by *selecting* possible candidates rather than by *constructing* them. The role of chance is to generate a wide range of possibilities. To speak more technically, we make a random selection from the space of possibilities and then check whether any can do the job asked of them.

To show how this might work out in practice Kauffman describes (1995, pp. 145–7) how one might find a molecule that mimics the hormone oestrogen. The process starts with about 1,000 molecules which one anticipates will include building blocks for the sought after molecule. These are then mixed into a solution of about 100 million antibody molecules. If the mixture is what is called *supercritical* then, over time, billions of new kinds of organic molecules will be formed, among which it is hoped that there will be at least one which will mimic oestrogen. Some oestrogen is now added to the 'soup' which already contains some radioactive oestrogen; this will have bound itself to any candidate molecules. If there are any such present, the added oestrogen will displace the radioactive variety and that occurrence can be detected. At that point it is known that there is at least one type of molecule present which can do the same job as oestrogen. The next step is to identify this molecule. Kauffman describes a method of repeated separation and dilution at the end of which it is possible to identify the successful molecule. Once it is identified, its structure can be determined and this paves the way for it to be manufactured in quantity. Chance enters this process when the supercritical mixture 'explodes' to produce the vast number of candidate molecules. The larger the number the greater is the probability that there will be one among them with the required properties. Chance achieves easily what it would be virtually impossible to do in any other way and it does so with an elegance and ease which should amaze us.

SAMPLING

Inference from samples is by far the largest and most important field for the deployment of chance to serve useful ends. It lies at the root of modern statistics and represents one of the great intellectual achievements of the last century. To many, it is not at all obvious that one can learn very much about the attitudes or voting intentions of a population of several millions by questioning only a few hundred people. In general, of course, one cannot but the possibility all turns on how the individuals are selected. A *simple random sample* provides one answer. At first sight it may appear paradoxical that the way to get hard information is to go down the uncertain route of chance but that is how it turns out.

First, we need to know what a simple random sample is or, more exactly, what a *simple random sampling process* is. To keep things on a level where we can easily imagine the sampling being carried out, imagine a school with 500 pupils and suppose we wish to know what proportion travel more than one mile to school. Because we do not have time or resources to question every individual, it is proposed that we base our estimate of this proportion on the answers given by the members of a sample of size twenty. Next imagine a listing all possible samples of this size. This is an enormously large number, 2.67×10^{35} to be exact, but as all of this is only going on in our imaginations, there is nothing to prevent us from thinking in terms of such numbers. We are only going to pick one sample and it must be one from this list because the list contains all the samples there could possibly be. So the question is: how shall we make the selection? The principle of simple random sampling says that each possible sample should have an equal chance of selection. The technicalities of how this is done need not detain us. We certainly do not need a full list from which one sample has to be selected because there are ready-made samples already available in tables of random numbers, or their equivalent which can be generated on a computer.

At first sight this may seem a rather risky way of trying to learn about the population of 500 pupils. Some samples will be very untypical of the population and if one of these happens to be drawn we shall get a very misleading picture. That is true but very unlikely, because most samples will be similar to the population as regards the proportion who travel more than a mile to school. All of this can be quantified by using the theory of probability. This shows that as the sample size increases the sample proportions become ever more closely clustered around the population value. Inverting this statement, the distance from the sample value to the population value will decrease and so our estimate of the latter will become ever more precise.

There is a lot more to statistical inference than taking fairly small samples from smallish populations. Much of it, in fact, concerns what are called infinite populations, where there is no limit to the size of sample which may be taken and where it makes sense to enquire what happens as the sample size becomes indefinitely large. None of this affects the basic ideas underlying the use of chance to learn about the world which is illustrated by the example given above.

The principal points which emerge from this discussion of inference from samples, echo those already familiar from the earlier sections of this chapter. One is the notion of fairness embodied in giving every possible sample an equal chance of being selected. It is notoriously difficult for an individual to pick a sample randomly from a small population of objects laid out before them. Simple random sampling provides a guarantee that there could not have been any introduction of bias, witting or unwitting. Secondly, there is the emergence of order from chaos as outcomes are aggregated. Although individual samples may show wide divergences from expectation, the sampling distribution, as it is called, shows a definite and repeatable pattern. There are near certainties on which we can base our inferences which emerge in the aggregate and which ensure that we almost always get things right in the long run. Although these things could be checked empirically, in principle at least, probability theory is available to save us the trouble.

THE ARTS

The element of surprise and unpredictability inherent in random processes has caught the attention of musicians and artists. Abstract creations in either field may, sometimes at least, take us beyond the conventional limits of imagination to explore unknown territory. To put it rather more grandly, chance enables us to explore the space of possibilities more thoroughly by opening up to us obscure corners which might otherwise be missed. One famous example of random composition goes back to Mozart.[4] In his *Musikalisches Würfelspiel* (K.516f, 1787) he wrote a piece for piano consisting of 32 bars – a 16-bar minuet and a 16-bar trio. Mozart composed 176 bars altogether from which the bars of the pieces were to be selected randomly. The original idea was that the bars would be selected by rolling dice. Today this can be done by drawing random numbers. In practice one can do this by visiting the web site http://sunsite.univie.ac.at/Mozart/dice/ and clicking on a button to play the piece. There are also options to choose instruments, one for each hand, and to make one's own random selection of bars. The number of possible musical pieces which can be constructed by this means is enormous – about 1.5×10^{15} for the minuet alone – so the chance of all the possibilities ever being heard is remote! The interest of the exercise is that pleasing and unexpected results can be obtained from an uncertain process.

SERIOUS GAMES

At the outset I reviewed the place of chance in games of the kind we play for amusement. There are more serious 'games'

---

[4] The material about Mozart was obtained from a note by David Bellhouse in the *International Statistical Newsletter* 30 (2006): 19–20.

in the worlds of economics, politics and management gener-ally.[5] The word *game* in this context is apt to hint at a lack of seriousness which may divert the reader from its real impor-tance. A game in this more general sense is a decision-making situation in the face of uncertainty, though here I shall inter-pret it more narrowly. For our present purpose a game is a situation where two or more parties, with conflicting goals, compete with one another for some desirable end. Compe-tition between firms for customers in the marketplace is an obvious example; candidates seeking the support of voters in an election is another. In the simplest case there are just two players with each trying to outwit the other. The uncertainty in such games arises mainly from lack of knowledge about what the opponent will do. Each player must therefore seek to see the game from their opponent's point of view, to try to anticipate what the other will do and to respond accordingly. Each player will have a number of strategies available to them and will have to guess what are available to the opponent. For each combination of strategies chosen by the contestants there is what is known as a *payoff*. This may not be monetary and its value may not be the same for the two players.

An extremely simple game, so simple that it would be rather boring to play, illustrates how it may be advantageous for a player to resort to chance in selecting a strategy. Each player – call them Bill and Ben – has a coin and has to choose which face to display. If they choose the same face, Bill wins and

---

[5] Although traces of the ideas go back to the early 1700s the mathematician John von Neumann created the field in a series of papers in 1928. His book with Oskar Morgenstern, entitled *The Theory of Games and Economic Behaviour* and published in 1944, was a landmark in the subject. However, in the early days it was difficult to find good applications in subjects such as economics to convince people of the theory's usefulness. More recently there has been a surge of interest extending into other fields such as biology.

pockets both coins. If they make different choices, Ben wins. There are thus four possible outcomes which may be set out in a table as follows:

		Ben	
		H	T
Bill	H	+1	−1
	T	−1	+1

The entries show the game from Bill's point of view; so a negative entry shows what Bill loses. If Bill and Ben both choose heads the gain to Bill is 1 and this appears in the top left position. If Ben chooses heads and Bill chooses tails the rules say that Ben wins, which means that Bill loses 1 as the bottom left entry in the table shows. If the game is played only once, each player will either lose or gain one unit and there is no way of telling, short of mind reading, which way it will go. The most that Bill or Ben can say is that their loss, in a single play, cannot be bigger than 1. However, if the game is played repeatedly, things become more interesting. There is now a strategy which ensures that either player can do better in the long run. Suppose Bill simply tosses the coin. When Ben chooses heads, Bill will gain 1 on half the time, when his coin falls heads, and lose 1 on the other half. These will balance out in the long run. The same will be true whenever Ben chooses tails. So Ben's choice is irrelevant to Bill because whatever it is, his gains and losses will balance out in the long run. Bill thus does better by ensuring that his gains never fall below an average of zero per play. It also turns out that he cannot do better than this whatever strategy he chooses.

The intriguing thing about this is that Bill does best by choosing heads or tails at random! Use of chance will yield more in the long run than can be achieved by long and careful thought. This seems paradoxical but it is not. The point is that if Bill did use some non-random choice strategy, then Ben would be able to observe his choices and, in principle, learn something about how Bill was making his choices. He could put this knowledge to good use by trying to predict Bill's choices. Knowing what Bill is likely to do enables Ben to take advantage of that knowledge. For example, if Ben thinks that Bill is likely to choose heads, he will counter by choosing tails and so win. The only way that Bill can ensure that Ben cannot predict his choice is to make his own sequence of choices unpredictable. And the only way he can be absolutely sure of this is by not knowing, himself, what he is going to do – by tossing the coin. Simple though it is, this example conveys the essential point I wish to make, namely that in a competitive situation it is important to deny information to the enemy. The sure way of doing this is to blind the enemy with chance.

GENETIC ALGORITHMS

Not only do humans use chance; 'nature' uses it also and on a grand scale. The whole evolutionary process has created living things exquisitely adapted to their environments. The combination of chance variation and natural selection has been a powerful creative force, fashioning the world as we know it. It is not surprising, therefore, to find that humans can be creative, albeit on a smaller scale, by imitating nature.

Genetic algorithms[6] are attempts to imitate nature by creating variation and selecting those products best fitted to survive.

---

[6] Genetic algorithms are an example of what are called computer-intensive techniques and for that reason did not become widely used until the dramatic increase in the power of desktop computing in the 1980s. They take their name from their analogy with genetic reproduction but, viewed mathematically, they are optimisation techniques which, to borrow a topographical metaphor, exploit the ability of computers to search systematically for a route to a summit.

The essentials are as follows. In nature the genome contains the code for reproduction. In successive generations the genome is modified by mutation and crossover. At each generation those most fitted to survive are best placed to breed and so form the next generation. Over a long period, the organism becomes better adapted to its environment by the accumulation of small beneficial changes.

The key idea which readily translates into engineering design, say, is to let chance generate a large number of variations on the current design from which the best can be selected. To do this we need some way of selecting the most successful examples to carry forward into the next generation. This is achieved by what is called an *objective function*. This provides a measure of how good a particular design is and so enables the best to be selected. The genome in this case is the design specification which lists every variable under the designer's control; it is the blueprint which would enable the design to be constructed. Typically this will be a list of, perhaps, hundreds or thousands of components, all of which could be varied. Adjusting them all in a systematic fashion to discover the optimum solution would be prohibitively expensive in time and resources. The speed and capacity of modern computers enables chance to step in and facilitate the exploration of a vast number of alternatives. Each is evaluated by the objective function and the best is selected as the starting point for the next round in the search for the optimum. *Blind* chance, as it is sometimes called, thus enables the optimisation process to *see* its way to a good solution. The true location of the *design* is in the mind of the creator of the algorithm who could not – and did not need to – see the details of the path which the process would follow. A human designer who took the alternative course of laboriously and systematically searching for the optimum would be quickly left behind.

This chapter may have appeared to be something of a digression with no immediate theological implications. However, its purpose has been to show that chance is not necessarily the enemy of all that is rational and that it is certainly not the antithesis of planning and design. Chance, in the shape of deliberate randomisation, has found an increasing range of applications in human activity both serious and recreational. This has been nourished and developed enormously by the fact that modern computers are good at performing large numbers of simple repetitive operations very quickly. If there are so many advantages in deploying chance in our own affairs, it is natural to ask whether there may not be benefit to God in achieving his own purposes in the wider creation. That is the subject of the next chapter.

CHAPTER 11

God's chance

In chapter 3 we saw that chance and order are not incompatible; many of the regularities in the world are the aggregate effects of chance happenings at a lower level. In chapter 10 we saw that such regularities enable us to use the lawfulness which they introduce for our own purposes. God, of course, created these possibilities in the first place, so it is natural to suppose that he would use the order which results to achieve his own ends.

In this chapter I show how this might have happened in three examples where the presence of a chance element has often seemed to rule out divine involvement. These are: the origin of a life-bearing planet, the evolution of life through natural selection, and the social order.

WHY MIGHT GOD USE CHANCE?

Is chance a real feature of our world and, if so, how can we reconcile its presence with belief in the God of Christian theology? That question was raised at the outset and it has overshadowed all that I have said since. I have already rejected the idea that what we see is only the appearance of chance – that in reality every single thing that happens is as a result of the direct and immediate action of God. This is logically possible, of course, but we have seen that, in practice, it is extremely difficult to mimic chance. Indeed, the only sure way of getting it right is to resort to a chance process itself. I have also rejected the idea that what appears to be chance mostly is chance, but that every now and again God intervenes at critical junctures to achieve his particular purposes. This view is equally impossible to reject solely on logical grounds, because such interventions need not be empirically detectable. There are also significant theological objections, to which we shall come later, but the central purpose of this book is to argue for the third view that God *uses* chance. In other words, that the chance we observe in nature is there because God intended it to be so. It serves his ends and furthermore, when properly interpreted, is conformable to a rational and biblical theology.

A clue as to the way we should proceed to establish this thesis will be to recall the advantages which the use of chance offered in some of the various human activities looked at in the last chapter. These may be listed under four headings as follows.

(1) Some random processes have outcomes which can be predicted with near certainty. In particular, their aggregate properties often obey simple laws which make things effectively deterministic at that level. If it is only the aggregate behaviour which matters, then the fact that there is a random substructure is of no significance.

(2) There is a principle of fairness, or equality, which is satisfied when selections are made at random. We saw this particularly in relation to random sampling but it is also prominent in random searches to ensure that 'no stone is left unturned'.

(3) A chance element in a system introduces a flexibility and resilience which makes it robust in the face of the uncertainties of the world.

(4) Randomisation often introduces the elements of surprise, novelty, flexibility and variety, which add immensely to the enjoyment of life and which develop a capacity to deal with the unexpected.

To the common way of thinking, chance, as I have already had occasion to point out, is synonymous with lack of purpose or direction and therefore hostile to belief in a God whose will and purpose is supposed to be expressed in the created order. My aim is to argue that this view is based on a total misunderstanding.

I aim to substantiate this claim by focussing, in turn, on three major issues where the apparent conflict between God and chance is most acute and where the theological implications seem to be most far-reaching: first, the existence in the universe of a habitable planet on which life has appeared; secondly, an evolutionary process in which chance plays an integral part, leading to the rich diversity of life on this planet including the production of sentient beings capable, for example, of writing and reading books; thirdly, the growth of human society in which a measure of genuine individual freedom seems to conflict with any overall divine purpose.

LIFE IN THE UNIVERSE

Here the gulf between what theology requires and science tells us seems to be wide and deep. The early picture was of the earth at the centre of things with humanity at the pinnacle of creation. The sun went inexorably round the earth, only stopping when it was commanded to do so by its Creator! Incarnation and redemption were comprehensible in terms of a small-scale universe with heaven above and earth below. Then the picture gradually changed. Copernicus placed the sun at the centre of the solar system and humankind was thus relegated to one of the smaller planets. Worse was to come in the discovery that our sun was one of a vast number of stars in a galaxy which seemed in no way special. The galaxies themselves were more numerous than the stars had once seemed and, furthermore, were receding at a rate that was hardly imaginable. From being a big fish in a small pond, humanity had become a mere speck located well off centre in a vast universe which seemed much too large to even notice it. It was hard to go on believing that humanity was as central to the meaning of things as the Bible seemed to suggest.[1]

This argument can, of course, be disputed on its own terms by asking why size, measured in light years or whatever, should be the relevant metric for judging these things. Or, why insignificance should be attributed to any particular location in a universe in which there is no natural point of reference.[2] If sheer complexity were the criterion for judging importance then the human brain, of which there are currently over 6 billion functioning on this planet, would provide good grounds for keeping the inhabitants of the earth in a central place. However, when we add to mere size the extremely hostile environment in which life had to gain a toehold, with extreme temperatures and bombardments from outer space, the last nail seems to be hammered in the coffin of the well-ordered, God-controlled earth which, in the pre-science era, had seemed so natural and credible.

The sheer size of the universe revealed by science has naturally encouraged speculation about the possibility of life on other planets. After all, we know for certain that life *can* occur, as our presence on earth proves without doubt. Why should there not be many other civilisations dotted around this vast universe? And what would that imply for doctrines such as

---

[1] The book by Primack and Abrams (2006) aims to tackle the seeming insignificance of humankind in the universe from a largely scientific point of view. Their main point has been noted in note 3 of chapter 1, but without adequate theological backing it does little to make the universe more 'human-friendly'.

[2] Paul Davies explains why the universe has no centre. See the section 'Where is the centre of the universe?' on page 27 of Davies (2006).

original sin and the need for redemption? It makes the whole Christian enterprise seem so parochial that one might scarcely credit that belief still exists among some, even, of the intelligentsia.

Scientific opinion is divided on whether we are alone, and there is precious little hard evidence to go on. If there were other advanced civilisations extant at the present time one might expect them to be indicating their presence, and large sums of money are being spent 'listening' for them – so far without hearing anything. In so far as probability arguments have been brought into play, the crudest goes somewhat as follows.[3] Given the size of the universe, there must be many stars with planets and some of them must have, or have had, or will have, conditions suitable for intelligent life to appear and evolve. The probability of life emerging at any location cannot be zero, because it has already occurred at least once, but given all that we know about the hazards involved the probability must be very small indeed. The expected number of such civilisations would then be obtained by multiplying the number of locations by the probability of life occurring at any one of them. (This assumes the probability is the same at all locations, but this assumption can be relaxed without affecting the point of the argument – see later.) This is where the trouble starts. The answer you get when multiplying a very

---

[3] There is a serious scientific side to the search for extraterrestrial life, and a growing subject called astrobiology. Another major interest is in how one would recognise alien messages were they to be directed towards our planet from living beings elsewhere in the universe. Conversely there is the question of what information should be transmitted from earth into space so that any recipients would recognise what they were receiving as coming from intelligent beings. The acronym SETI (Search for Extraterrestrial Intelligence) is often used to describe the community working in this field. The public imagination goes much further, and many people, in America especially, will testify to having been abducted by visiting aliens.

small number, the probability of life, by a very large number, the number of planets, can be almost anything; it depends on knowing the magnitudes of the two numbers fairly precisely. Yet, if the approach is to be of any use, we need to know whether the answer is much less than one, in which case life elsewhere is unlikely, or much larger than one, in which case the opposite is true.

Conway Morris (2003, pp. 229ff.) quotes a number of examples of this kind.[4] He refers to Gaylord Simpson, for example, who agreed that there were many locations at which life might have occurred and that the probability at any one must have been very small. From this he claims that 'the Universe is a big place, so however uncommon life was, the total number of planets with life must be quite large' (p. 229). There is no *must* about it. Everything depends on just how many planets there are and what the probability actually is. George Beadle (p. 230) was another whose arguments appear to depend on the fact that numbers may be very large and probabilities very small without an adequate logic for relating the two.

A slightly more sophisticated attempt to solve the problem is embodied in Drake's formula.[5] All this does is to break the problem down into smaller components, without

---

[4] The page references in this paragraph are to the beginning of his chapter 9 entitled 'The non-prevalence of humanoids'. George Gaylord Simpson was a notable evolutionary biologist of the last century. In the notes to chapter 9 Conway Morris gives more information about his work. George Wells Beadle was an American geneticist who shared the Nobel prize in 1958 with Edward Lawrie Tatum for work showing that genes act by regulating the chemical events within the cell.

[5] Frank Drake, an American astronomer, formulated what is known as Drake's formula, or equation, in 1961. It exists in more elaborate forms than that implied by the discussion here but, in any version, it is essentially a tautology. Its purpose is to break down the problem of finding the probability into more manageable parts. It tells us something about the structure of the probability but nothing about its value.

materially contributing to its solution. In fact it has been shrewdly remarked that it is merely 'a way of compressing a large amount of ignorance into a small space'. This quotation is reproduced in a paper by Lineweaver and Davis (2002, p. 11) which does take matters a little further. Drake's formula represents the proportion of stars in our galaxy orbited by planets that have had independent biogenesis as the product of three proportions. These are: the proportion of stars in our galaxy with planetary systems; the proportion of those planetary systems that have a terrestrial planet suitable for life in the same way as the earth; and the proportion of these suitable planets on which biogenesis has occurred. We do not know any of these proportions, but progress might be made if some estimates could be made of one or other of the components. Lineweaver and Davis claim that they have a method which tells us something about the last of the three proportions.

The idea is quite simple. On an earth-like planet there will be a 'window of opportunity' when conditions are propitious for the emergence of life. If it is easy for life to emerge it is more likely to do so early in that period than if it is hard. By establishing that life appeared on earth at a relatively early stage, we can conclude that the Earth's environment is in the easy category, and hence that life will have arisen quite frequently on other, 'similar' planets. By this means, and by making further assumptions, the authors claim that the probability of life arising under these conditions is likely to be greater than 13 per cent. Hence there is a fair chance that biogenesis will have arisen once a suitable planet occurs. This interesting result still leaves us a long way short of an estimate of the proportion of life-bearing planets in the galaxy.

There is, in fact, a more relevant calculation which can be made which shows the whole issue in a totally different light. Its novelty lies not in that it replaces the naturalism of the approach just discussed by a theological perspective, but rather that it retains all of that without excluding God. We shall consider how God might use these same natural processes to achieve a life-bearing planet. In pursuing this strategy we shall adopt what may seem to be a presumptuous usurping of God's place by asking what God might do. However, we often do this unconsciously and it is better to be open about it than to adopt a false deference. Essentially we are using our imaginations to contemplate a set of possible worlds.

ONE WORLD OR AT LEAST ONE WORLD?

In this chapter I am asking how God might use chance, and at this point, how he might use it to bring planets into being capable of supporting intelligent life. The deity would begin by recognising that, in the sort of universe brought into being by the 'big bang', the chance that such life might arise at any location was bound to be extremely small. The universe, therefore, would have to be very large for there to be any chance of its happening. But how large? Large enough for there to be a very high chance of its happening *at least once*. At this point a part of probability theory becomes relevant. However big the universe is, the actual number of life-supporting planets will not be certain but it will have a probability distribution. Furthermore, and surprisingly perhaps, we can deduce what the form of that probability distribution will be. If we have an exceedingly large number of locations and an exceedingly small chance that any of them will produce life, then we know that the distribution will have the Poisson form, a distribution we have already met in chapter 3. This does not require the probabilities to be equal – just small.[6] There is one further assumption which is that there will be independence between what happens in one part of a galaxy from what happens in another. Given the size of the universe, with such immense distances between stars and galaxies, there seems to be no problem about that. The form of the Poisson distribution depends solely on the (unknown) average number of life-supporting planets. Taking all this into consideration, we can easily determine what this mean number has to be in order to ensure that life occurs at least once. This means that once God has decided what the probability of getting at least one occurrence should be, the average number of occurrences and the distribution of occurrences is determined. Let us put some numbers on these abstract statements to see just what is involved.

Suppose that the Deity would be satisfied with a probability of achieving at least one 'success' equal to 999/1000. A simple calculation shows that this can be achieved with an average number of 5.9 life-bearing planets. Interestingly, this makes the chance of getting exactly one occurrence a mere 0.016, so our ensuring that we get at least one means we shall very likely have a lot more than one. A more risk-averse Deity who wanted to raise the chance to 999,999/1,000,000 would get an average of about fourteen. These simple calculations have far-reaching consequences. It appears that the number of inhabited planets would be relatively small if such a method of creation were used. Furthermore, since any given planet would only be able to support intelligent life for a relatively short time, and given the fact that the appearance of life would

---

[6] This is a standard result in probability theory which will be found in most texts on the subject. The more elementary version, which supposes that the probabilities are both small and *equal*, is usually found as a limiting case of the binomial distribution, but the generalisation to small and *unequal* probabilities is straightforward.

be spread over a long period, it would seem unlikely that there would be more than a very few civilisations extant at any one time. It is reasonable then, that a planet such as ours has had no communication from elsewhere in the universe. Although this is a very slender thread to hang anything by, we do at least note that our experience is consistent with this theory.

It is also clear that there is no way one could expect to get *exactly* one world. This method of creation, if it is to yield at *least* one, must almost certainly yield several. It is also clear that there is no need to have more than a handful if the aim is to get at least one. What we observe is thus consistent with the hypothesis that all God had to do was to make the universe large enough and then leave things to chance. This line of argument might seem to have pushed God to the sidelines and accorded him a marginal and unskilled role, which is surely inconsistent with the power and majesty of the Christian God. Should he not be at the centre of a detailed and carefully constructed planning enterprise and in close control of the whole process? By no means. If there is a simple and elegant route to the creation of life-bearing planets ought not God to know about it, and is it not a mark of his greatness if he chooses to use it? His thoughts are not our thoughts, and perhaps this analysis gives us a hint of how his ways might not be those which immediately occur to us.

CHANCE IN EVOLUTION

Nowhere is the possible role of chance more hotly contested than in evolution.[7] Darwin's theory of evolution by natural selection, in which random mutations provide the variation on

---

[7] The perceived conflict between evolution and creation rears its head at many junctures, in this book as elsewhere. A good treatment from a scientific and Christian point of view is provided by K. B. Miller (2003); this is an edited volume which is particularly suitable for students and others from

which selection works, is seen by some as the very antithesis of the purposeful direction which seems to be the very essence of Christian theology. The battle lines[8] are drawn most sharply in the United States of America, where evolution is often seen

a more conservative Christian background. A gentler attempt to woo a similar audience is given in the first part of Colling (2004). Another very useful source is K. R. Miller (1999). In an entirely different vein, those who wish to explore the phenomenon of creationism will find Coleman and Carlin (2004) very informative. I know of no work comparable to these treatments which provides a convincing case for the creationist side of the argument.

[8] The lines are not as sharply drawn as this rather bald statement might suggest. Many who are theologically conservative are fully committed evolutionists. In the preface to the book edited by K. B. Miller (2003) for example, Miller writes that 'having become deeply frustrated with the often fruitless and divisive nature of the "creation/evolution debate" within the evangelical Christian community, I hope to move the conversation in a more positive direction'. The tendency within this collection is to regard chance as due to ignorance. Miller speaks approvingly of 'some theologians who see God's action exercised through determining the indeterminacies of natural processes' (p. 9). The tension becomes acute for those like Terry Gray, who expresses it in his section on theological musings on pages 385–6. He concludes that 'while we may not be able to resolve the apparent contradiction, using human reason, we readily affirm both truths'. One cannot help wondering whether part of his trouble arises from regarding the Westminster Shorter Catechism as the definitive interpretation of scripture.

There are others, such as Morton and Simons (2003), who do believe that God uses chance and who find biblical support for the idea. However, they also believe that God controls chance, but in what sense is not clear.

There is a similar lack of clarity in Garon (2006). In chapter 19, entitled 'Does chance have dignity?', he comes close to the idea of seeing chance as being within rather than outside God's will. He recognises that disorder can serve a purpose as when mixing fruit into a cake – we want it disordered. However, he still seems to be saying that God knows it all. On page 14 he says, 'And, being infinite, the Creator knows the outcome of all things for all time. Nothing can be accidental in the sight of the Lord. What we call "Accidental outcome" is accidental only to ourselves, not to God.'

In a very recent article Woolley (2006) has argued that the Anglican theologian Leonard Hodgson (1889–1960) anticipated the idea that chance is used by God. This is a matter which, clearly, needs to be pursued.

as necessarily atheistic by ruling out the notion of design, and hence, of a Designer. The idea advocated here, that chance is part of the creative process, is not even entertained. One gets the impression from Denyse O'Leary's *Design or Chance* (2004) that the real battle lines are drawn between atheistic Darwinians on the one hand and Intelligent Design advocates on the other, with theistic evolutionists and young-earth Creationists playing a secondary role. Theistic evolutionists are seen by her as being 'Darwinists with a slight glow of faith' (p. 240), their problem being to distinguish themselves from the atheistic Darwinians.

I have discussed and dismissed Intelligent Design, which seeks to eliminate chance altogether, in chapter 7 and that leaves us with evolution in some form as the only credible alternative. The issue to be considered here is whether chance-driven evolution can be seen as part of God's creative process, expressing his purpose, or whether, as some Darwinians believe, it is a blind and undirected process.

Richard Dawkins is prominent among the latter and he would certainly not see chance in evolution as expressing any purpose whatsoever. But, oddly enough, his argument does support God's involvement but at a different level. Dawkins' belief that chance is sufficient is a continuing theme in his writings. His invention of computer-generated *biomorphs* to illustrate the idea was discussed in Bartholomew (1996, p. 177). However, a more recent book (Dawkins 2006 [1996]) has introduced the metaphor of climbing Mount Improbable to illustrate the point. The mountain has one precipitous face which is extremely difficult to climb and a reverse side which has an easier gradient. Ascents on this side may be longer and more winding but are less arduous. Constructing complicated organisms is like climbing the mountain to reach a summit. The chance of doing it by the steep face is extremely small but by using the easier route there is a much higher probability of success.

To see how this works out, let us recapitulate the essentials of Dawkins' case. He starts by pointing to examples in nature of extremely complicated constructions which seem to demand design as part of their explanation. The eye and the haemoglobin molecule are just two examples. He even quotes, to their disadvantage, distinguished scientists who have declared publicly that they could not believe that such things could have arisen by chance. Dawkins agrees that they would certainly not have arisen if their creation was merely by collecting the parts and assembling them 'at random'. But, he goes on to say that that is not how they came about. Chance did, indeed, produce the variation on which natural selection operated but, instead of climbing the impossibly steep cliff of Mount Improbable, evolution took the roundabout route on the easier slopes on the other side of the mountain. In short, the extraordinarily small probabilities which are quoted are irrelevant; the true probabilities are much larger and hence the evolution of complex organisms and molecules was not so unlikely after all.

By correctly challenging the very small probabilities which so many, du Noüy and Overman among them, have seen as providing incontrovertible evidence for design, Dawkins supposes that he has demolished the case for a designer God behind the process. On the contrary, he has helped to make credible an evolutionary process which is capable of producing the complexity in the world and hence, as I have argued, that it could be the God-chosen way of creating things. I also have been arguing that the inferences to God based on very small probabilities are often invalid. Only by first doing that, is the way clear for a much deeper and more subtle account of what has been going on.

Dawkins, it should be noted, establishes that there may be a way to the peaks which makes the evolution of complex things possible. He does not actually show whether a route exists in particular cases. Is there a route, for example, which leads to human-like creatures? In attempting to decide this issue, it is relevant to note that there is a difference, among scientists, about how evolution develops, and one view is very much more friendly to theism than the other.

The alternative positions are set out in Gould's *Wonderful Life* (1989) and Conway Morris' *Life's Solution* (2003). Gould created many metaphors to convey scientific ideas to his readers. One of the best of these, and often quoted, is that of rerunning the film of evolution. According to Gould this would never end up at the same place twice. This seemed so obvious, and at the same time, so threatening to religious belief, that if the destination reached is the result of a 'random walk', how can it possibly express any purpose? Conway Morris, an evolutionary palaeobiologist who also worked on the Burgess shales from which Gould drew so much of his material, reached a different conclusion. Their opposing conclusions are best summarised in their own words, beginning with Gould. Thus: 'This book is about the nature of history and the overwhelming improbability of human evolution under themes of contingency and the metaphor of replaying life's tape' (1989, p. 51) and 'Rerun the tape of life as often as you like, and the end result will be much the same' (p. 282). Gould thought that, in the reruns, evolutionary history would never repeat itself because chance was brought into play at every juncture.

Conway Morris, on the other hand, built his case around the notion of *convergence*, according to which evolution is liable to produce similar solutions to similar problems wherever and whenever they occur. Towards the end of *Life's Solution* he says: 'The principal aim of this book has been to show that the constraints of evolution and the ubiquity of convergence make the emergence of something like ourselves a near-inevitability. Contrary to received wisdom and the prevailing ethos of despair, the contingencies of biological history will make no long term difference to the outcome' (2003, p. 328).

Their interpretations imply radically different probabilities for much the same outcomes. One or other of them must be mistaken. If Conway Morris is nearer to the truth it is perfectly credible to suppose the chance and necessity of classical Darwinism is sufficient to produce human life. If that is so, it makes sense to suppose that God designed it that way. It makes sense, also, to see God using chance to achieve a desired end.

We do not have to choose between equally well-supported scientific cases because, as noted in chapter 5, there is a fallacy in the probability reasoning which Gould and others have employed. Let us remind ourselves of Gould's argument. He begins *Wonderful Life* by debunking the idea that evolution has been an inevitable progression by what he calls either a ladder or a tree. In opposition to both of these pictures he wishes to give full priority to chance, or contingency as he often calls it. If we think of evolutionary paths as being possible routes along which life might have proceeded, there will be a very large number of branch points at which the path could have taken one course or another. The probability of completing any particular path is obtained by multiplying together the probabilities associated with each junction, and if the choices are independent, the product will be extremely small, so one could assert that any repetition would almost certainly yield a different route. I argued that the assumption of independence was false and so the conclusion does not follow. Nevertheless, there is one further thing which should have given Gould pause for thought. If all the choices are indeed taken at random, perhaps Gould should have noted that the particular path which was realised is the one which leads precisely to the end-point which we occupy as sentient beings able to ask these questions. If an event having an exceedingly small probability actually occurs, we should at least ask whether that puts us in a position to reject the hypothesis – namely of randomness – on which it was calculated. I have already discussed in chapter 6 the logic of rejecting hypotheses that have very small probability but shall not pursue it here because Conway Morris' objection to Gould's argument is more fundamental. Gould does not appear to have considered the matter of independence at all, but Conway Morris did notice the problem by recognising the ubiquity of *convergence* in evolution. Convergence refers to the coming together of different evolutionary paths. Gould supposed that this did not happen and that each path reached an end-point which was distinct from all others. Conway Morris' book contains an extensive catalogue of examples showing that this is not actually the case.

Before going into this I need to be a little more explicit about what is meant by an evolutionary path. Such a path is a full description of the biosphere as it has developed over time, with any point on that path representing its state at a particular time. Convergence occurs if several possible paths have things in common. For example, small mammals living underground live in similar environments and these exert selective pressures to evolve similar characteristics – for communicating, say. Thus paths which hitherto had been quite separate, at a later point may converge in this respect. Convergence may refer to things as diverse as anatomical structure or social organisation. The important thing is that the environment in which they live poses challenges for the organism which are met in much the same way. Usually we think of natural selection as favouring the fittest in terms of their ability to survive and thrive. In bringing convergence into the picture we are recognising the fact that the physical environment, including the presence of other creatures, also plays an important part in who survives.

What matters, of course, is that humans or some similar creatures should have recognisably human characteristics and functions. According to Gould, and those who think like him, the various paths will be so different that the appearance of humans in our particular path would be purely fortuitous. Conway Morris claims that the empirical evidence points in the other direction. He thinks that the evidence shows that the physical and other constraints are such that humans, of some kind, are very likely to have turned up in our world or in virtually any other universe, for that matter.

The suspicions raised by Gould's assumption that pure randomness determines the path of evolution have been amply confirmed by Conway Morris' impressive empirical evidence. Things are very much more complicated than Gould supposed. Conway Morris has not produced a tested probability model to show that human life is an almost inevitable consequence of a universe such as ours, but at any rate he has demolished once and for all the attractive but flawed model of contingent evolution.

It is possible to take a different line and to argue that evolution by mutation and natural selection is God-guided, without invoking the empirical support provided by evolutionary convergence, and Keith Ward has done so in *God, Chance and Necessity* (1996). Starting from essentially the same point as Gould, Ward recognises that natural selection alone would make the emergence of human life extremely improbable but, as a theist, he sees no problem in supposing that God 'weights the probabilities' in some way to ensure that evolution reaches the desired destination. This might be by guiding the selection of mutations in some way, though it must be remembered that most mutations are harmful, in any case. The logic of Ward's argument is unexceptional. If there is a God and if he desires a human outcome to evolution and if he has power to influence natural processes in the world, as Ward claims he must, and if chance and necessity are insufficient by themselves then God will surely act to do what is necessary. Ward puts it as follows:

> It would be entirely reasonable to conclude that the hypothesis of God – of a cosmic mind which sets up the processes of mutation so that they will lead to the existence of communities of conscious agents – is a much better hypothesis than that of natural selection. For on the God hypothesis, the development of sentient life-forms from simple organic molecules is very highly probable; whereas on the natural selection hypothesis, such development is very highly improbable.
>
> It is open to a theist to argue that natural selection plus the basic laws of physics does make the development of sentient life-forms probable. In that case, one could hold that God has designed the basic laws so that, in the long run, in one way or another, conscious beings would come to exist. One would see natural selection as the way in which God works, without interference in the laws of nature, to realise the divine purposes in creation. God would not be needed to explain why natural selection moves in the direction it does, when it could easily have moved in some other direction (or in no direction at all). But God would still have an explanatory role, in providing a reason why this set of physical laws exists, and in assigning a goal (of conscious relationship to God) to the process of evolution.
>
> For such a view, God and natural selection would not be competing hypotheses. God would be the ultimate cause of the finite causal processes embodied in natural selection, but would not interfere in those finite causal processes, as an additional cause. I am not happy with accepting this otherwise attractive view. I am not convinced that the principle of natural selection alone makes the emergence of rational beings probable. (1996, p. 76)

Ward was writing without taking account of convergence which, according to Conway Morris, makes the evolution of sentient beings very much more likely. It is thus not necessary to invoke the direct action of God at all.

Gregerson (1998)[9] is another theologian who has looked in a constructive way at the role of chance in the creative process. He has introduced the term *autopoietic process* to describe the way in which physical processes may develop using a God-given propensity to move towards a God-determined end. This is conceived as something less than full self-determination but it still retains a role for God's ongoing involvement. This is conceived of as an influencing of the probabilities rather than the outcomes themselves. God thus sustains and directs the process without its requiring his detailed involvement. As with Ward, this view presupposes that chance alone is not sufficient to achieve the desired end.

To some this indirect participation by God might seem like a step backwards and to play into the hands of those Darwinians who see no need for divine involvement at all. This is just what advocates of Intelligent Design fear, and what they are concerned to exclude by insisting on the necessity of a Designer. However, evolutionary theists have a powerful riposte to this move which is to agree that the natural world does, indeed, show evidence of design but that its explanation is much more remarkable. The God whom they worship not only designed things the way they are but also designed and brought into being the 'machinery' for making them! There is no need to leave a glaring gap between the conception in the mind of God and its execution in the world. The act of creation is an elegant and, relatively speaking, simple way of doing it all at once.

[9] There were responses to Gregerson's article by Rudolf Brun, by Richard McClelland and Robert J. Deltet, and by Langdon Gilkey in *Zygon* 34 (1999): 93–115. Gregerson replied in the same issue of the journal (pp. 117–38).

Ward is thus half right when he says:

> There is no physical mechanism that produces such a bias. Yet it is not left entirely to chance. It is the being of God, which alone sustains the universe, that exercises a constant causal 'top-down' influence on the processes of mutation and natural selection, and guides them towards generating sentience and the creation and apprehension of intrinsic value. (1996, p. 83)

There is no physical mechanism – there does not need to be – but chance and necessity alone are sufficient to do the job in exactly the way God intended.

THE SOCIAL ORDER

Here I make an abrupt turn into the social world, not made up of mindless elementary particles, but of sentient beings with minds of their own. How can chance possibly arise in this world, and if it could, what use could God possibly make of it? The relationship of choice and chance is the subject of chapter 13. My intention here is to make a link between the 'order out of chaos' of chapter 3 and the 'choice and chance' of chapter 13 which falls into the present discussion of how chance can be used to attain desired ends. There is, in fact, an important link between systems of particles behaving randomly and systems of people exercising free choices. This link is the lack of control of what is going on at the individual level.

In the first section of this chapter, where we were looking at the generation of life-bearing planets, we did not assume any immediate divine control over the natural processes out of which planets arose. In the second section the focus moved to the process of natural selection driven by random mutations which, considered individually, expressed no purpose. Here we have collections of people expressing their own individual purposes independently of any overall direction. What all these systems have in common is that there is no control at the micro or individual level in the case of human society. The fundamental question is then whether the overall Designer of those systems, having no detailed control, can nevertheless still achieve global purposes. If so, and we have seen that it is so, there are enormous advantages to be gained. In the evolution, whether of planets or people, there is an economy and elegance in the 'production process' which speaks eloquently of the character of the Creator. In human affairs, global purposes can be achieved at the same time as allowing individuals their freedom. This is a truly remarkable situation, the significance of which seems to have been largely lost on theologians.

We now come down to particulars. In chapter 3 we had a hint of how collections of people acting independently might exhibit some of the aggregate regularities we might otherwise associate with collections of particles. When we move on to choice and chance, in chapter 13, the similarities will become more pronounced. Florence Nightingale was interested in aggregate social laws though, as we saw, she thought they were divinely decreed in a very direct manner. There are, however, many regularities in aggregate human behaviour which result from something very like chance at the individual level. If that is so it follows that God would know that free will at the individual level led to predictable consequences at a higher level. Allowing free will into the scheme of things is not, therefore, as risky as it might seem. Individual free will may have benefits to the individual – valuable in themselves – but also have secondary benefits by achieving God's will at a higher level of organisation in society.

An example[10] may help to make the point. In many societies a person is free to leave their employment whenever they wish, subject, perhaps, to giving appropriate notice. This is one way in which individuals can exercise their personal freedom, yet a remarkable pattern emerges when we look at the frequency distribution of lengths of service of many people in the same type of job. The form of that distribution is not the main point here but its shape is known as lognormal. Typically there are a large number of relatively short lengths of service and fewer longer lengths, with the frequencies tailing off into the very long lengths of service. This means that it is possible to plan recruitment and suchlike on the large scale with considerable confidence. There is thus no reason to suppose that God would not take account of such regularities in considering the global characteristics of social affairs. It is tempting to go on and speculate on whether history, which is the sum total of human activity, is determined rather as the gas laws are, for example. Is it true that the free choices of individuals acting at the individual level determine the general course of history? Doubtless there are some broad features such as the relationship between decisions to leave a job, and the general characteristics of the labour market, which may indeed seem to support that conclusion. However, such a conclusion would be premature because there is one very big difference between physical systems and social systems.[11] This is the matter of size. In any of the physical systems we have considered there

[10] This is one of the oldest and best-known examples of a law-like phenomenon in the social sciences. It goes back at least to 1955 and has been widely used since in what is now called human-resource planning. See also the following note.

[11] Adolphe Quetelet was the pioneer in the field of the scientific study of aggregate social phenomena (see above, chapter 8, note 2). The author's own contribution to this field is contained in Bartholomew 1982. Quetelet coined the term 'Social Physics' and the similarity between the aggregate

may be billions or trillions of particles and it is in the regularities they display that the aggregate order is to be found. Social systems, on the other hand, may sometimes consist of millions of people but thousands or, even hundreds are more common. In such relatively small systems the emerging regularities will be clouded by the uncertainties resulting from the smaller numbers. There will, therefore, be a less clear-cut boundary between the micro and macro aspects of the system. Nevertheless, enough order may be retained over the global behaviour to justify the benefits which freedom of choice for individuals confers.

In this example it may be more accurate to speak of God *allowing* rather than *using* the uncertainties but the end effect is much the same.

behaviour of collections of particles and human populations has been noticed by other physicists. Social scientists have been somewhat lukewarm about this and many would share Steve Fuller's opinion that 'Social Physics can never explain complex decision-making processes such as choosing whom to marry' (*New Scientist*, 4 June 2005). Few physicists would claim that it could!

CHAPTER 12

The challenge to chance

The fact that God *could* do things through the medium of chance does not mean, of course, that he actually does so. The view expounded in the last chapter is challenged by those who take a strong view of the sovereignty of God. One of the clearest attacks comes from John Byl, who claims that to introduce ontological chance is scientifically unwarranted, philosophically objectionable and theologically inconsistent with the sovereignty of God. This argument is met by attempting to overturn each contention, and in particular, by arguing that a sovereign God does not need to be directly involved in those matters to whose outcomes he is indifferent.

BACK TO GOD'S SOVEREIGNTY

It is one thing to argue that God *could* have created the world in a manner which allows chance a real and important role. It is another matter entirely to argue that he did actually do it in that way. There are many things that we *can* do but there are some which, our friends would argue, we would not have done on the grounds that it would simply be 'out of character' for us to behave in that way. Maybe it would also be out of character for God to act as I proposed in the last chapter. After all, there is nothing in what we observe to show that what looks like chance must be chance. In spite of the complexities involved in mimicking chance, it is not impossible to do so. The objection to what I have been arguing comes from theology, not science. Some, such as Sproul and Overman, have left us in no doubt about their view of the matter. A more closely reasoned objection is provided by John Byl (2003) and I shall make his article the point of departure for a defence of the proposition that chance has its origin in God. There are two stages to the argument. First, I take the objections which Byl raises to the idea that chance has a place in God's scheme of things. Secondly, in answer to Byl, I aim to strengthen the case by noting that there are positive theological gains in the added flexibility that an uncertain world allows.

BYL'S OBJECTIONS

The central thesis of this book is that chance plays a positive role in the world and that it does not undermine God's sovereignty. John Byl has provided a sustained critique of this position and his article provides a convenient basis for examining the implication of his position in detail.

At the outset a possible source of misunderstanding, about what Byl calls ontological chance, should be removed. This is essentially the same as what I have been calling pure chance. The term refers to happenings[1] for which no cause can, even in principle, be found. That is, there is nothing we can observe which has any predictive value whatsoever. Radioactive emissions are the archetypal example. It should be made clear that it is not a necessary part of my thesis that ontological, or pure, chance is a *necessary* ingredient of the created order. Rather, that chance is present and plays a role which is *indistinguishable* from pure chance. Whether or not there are actually any events which are uncaused can be left as an open question,

---

[1] Byl's point, of course is that there are no such happenings – see the following note. Here we are merely clarifying how such happenings are defined.

but I certainly do not to wish to rule out the possibility of ontological chance at the beginning. The proposition I wish to defend is that the world may be treated *as if* the uncertainties are due to the operation of pure chance. How God might engineer chance happenings is a separate question. But, for the moment, I shall engage with Byl on his own ground.

Byl's argument is that ontological indeterminism 'is scientifically unwarranted, philosophically objectionable and theologically inconsistent with a strong view of divine sovereignty and providence'. I shall consider each of these objections in turn.

First, the claim that ontological indeterminism is scientifically unwarranted. In Byl's view this is because there can be no empirical evidence to settle the question of whether ontological indeterminism exists. Even if, for example, radioactive disintegration has all the marks of pure indeterminacy, we do not have to believe, Byl argues, that the radioactive events we observe are actually uncaused. He claims that to make such an assertion we would have to establish that there was nothing whatsoever in the observable world having any predictive value. To observe the present or past state of the world in the detail needed to establish this is clearly impossible, so we can never be sure that we have not overlooked some feature which plays a causative role. Essentially the argument is therefore about the lack of empirical evidence for asserting that there are uncaused events. Once this is accepted, Byl has a logically secure position but it does not follow that this makes the opposing position 'scientifically unwarranted'. His view has become increasingly exposed to the growing empirical evidence that ontological indeterminism provides the simplest, if not the only explanation of much that we observe. It is true that the question can never be settled with absolute certainty but there *can be* an accumulation of empirical evidence all pointing in one direction. A hypothesis which accounts for the indirect empirical evidence has to be taken seriously, even if it can never be supported by direct evidence. In summary, Byl's argument is not, in my view, sufficient to justify the claim that the hypothesis of pure chance is scientifically unwarranted.

Secondly, the philosophical objection comes from the notion of lack of cause. Byl believes that the principle of sufficient reason[2] requires that there *always* has to be a cause. If there is no *physical* cause, then some other agent, human or divine, must be responsible. This leads him to the position that a sovereign God must be in absolute control of everything that happens and, in particular, of determining events at the quantum level. Byl is aware of Bell's inequality and the experimental evidence which it provides for the pure randomness of quantum events. Bell's[3] work places severe limitations on the so-called 'hidden variable' explanations of quantum behaviour which would allow the deterministic interpretation which Byl requires. In particular, hidden-variable theories violate the property of classical physical theories which forbid caused effects at a distance. However, Byl is prepared to entertain such counterintuitive possibilities in order to retain causality. Again his position can be maintained, but only as the ground on which he stands is gradually eroded around him.

---

[2] The principle of sufficient reason says that anything that happens does so for a definite reason. It is commonly attributed to Leibniz and was elaborated by Schopenhauer into four forms. To invoke the principle in the present case is to do little more than say that there must be a cause for every quantum event. If that is the case, then it is obviously philosophically objectionable to suppose otherwise.

[3] John Bell (1928/90) was a Northern Irish physicist who proved a theorem relevant to the interpretation of quantum mechanics. This result has been described as the most profound discovery of science. Bell's inequality is derived from the theorem and this makes it possible to make an empirical test of whether deterministic theories of quantum behaviour are possible. A considerable body of empirical evidence has accumulated to show that they are not.

Thirdly, and this is the theological objection, the strong view of the sovereignty of God requires that God knows everything and is in detailed control of every happening. This is in opposition to the view expressed by Peacocke, Ward and others,[4] including the present author, that God might willingly give up some detailed control in order for a greater good to be realised. I have argued elsewhere in this book (chapter 1, p. 15 and chapter 11) that the strong view of God's sovereignty is not as well-defined as this simple account suggests. Since randomness at one level implies order at another level then, if one expects the order at this second level to express purpose, one cannot have it without randomness at the lower level. We cannot have 'simultaneous sovereignty', if we can so describe it, at both levels. This makes it difficult, if not impossible, to say exactly what the strong form of sovereignty actually means. That, in turn, makes it very difficult to know what the doctrine of God's sovereignty asserts about the world as we now know it to be.

The most interesting aspect of Byl's critique arises not in relation to quantum events, but in relation to human decision making. Human freedom seems to require a degree of indeterminism in the created order without which it becomes a mirage. Hence, others have argued, the development of human freedom requires there to be sufficient space for that freedom to be exercised. Chance seems to provide just the flexibility required and therefore to be a precondition of free will. Byl argues that this is not necessary and he opts for a compatabilist[5] view

---

[4] See, for example, Peacocke (1993, pp. 152ff.), Ward (1996, especially pp. 8off.) and the author's *God of Chance* (Bartholomew 1984).

[5] Compatibilism is the philosophical doctrine which says that humans can be free, in a real sense, in a deterministic world. In essence, individuals will freely choose to do what it is determined that they shall do. In this way humans can be held morally responsible for their actions.

of free will, which supposes that the choices which individuals freely make are compatible with what God wills. All the uncertainty is, therefore, on the human side.

At first sight, however, the alternative idea of genuine human freedom makes things worse rather than better. If we interpret it to mean that human beings act purely at random, this is hardly what we would expect from responsible human adults. The notion of free will is surely not the same thing as 'blind' will. Yet it is empirically true, as noted in chapter 8, that many human activities do sometimes appear to be random, at least in the sense that when observed in the aggregate, behaviour appears to be exactly what one would have obtained if decisions were made by some random process.

Before rejecting this idea out of hand, we must examine it more carefully. It is here that my definition of pure chance comes to the rescue, because I defined it simply in terms of unpredictability. Is it really true that human choices are, to some degree, completely unpredictable – that is, there is nothing we could observe about the individual or the world which would be of the slightest use in explaining that person's actions?

I shall now argue that the position of a human choice maker is fundamentally different from physical entities such as the electron. A human decision *can be* rational and yet appear to the outsider *as if* it were made at random. The reason is that each individual has an *internal* mental world as well as an *external* world. To appear random to the outsider, choices have to be unpredictable by reference to any feature of the external world because that is all they can observe. To appear rational to the choice-maker they have to depend – and may only depend – on features of the internal *and* external worlds. This internal world consists of ideas, reflections, beliefs, attitudes, perceptions, and so on, which may have been formed under the influence of the external world but are not fully and publicly observable in that world. If, of course, attitudes, beliefs and opinions are elicited by someone in the external world, they then become part of that world and could be used in predicting behaviour. But how those internal factors are used by the individual decision-maker is unobservable and yet may be totally rational.

In summary, therefore, I am proposing that there is a purely random element in human decision making, but only in the sense that choices depend, in part at least, on inherently unobservable factors known only to the choice maker. This ensures that they will appear unpredictable to the external observer because there is nothing available to the external observer which has predictive value.

But what is the 'self' residing in our person that processes this information, private and public, to arrive at a decision? It seems essential to require that there is a self which depends upon, but supervenes[6] upon the body of experience which each of us finds in our internal world. It does not preclude the possibility, of course, that the divine being should interact with that self and thus be a party to our choices. However, this interaction must be voluntarily restricted on the divine side to avoid taking away the very freedom which the internal world is identified to allow.

My conclusion, therefore, is that human choices are real and they may have the appearance of pure randomness because the causative factors behind them are part of the internal, private, world.

Byl goes considerably further than this in the matter of human choice. I have recognised that individual decisions may

---

[6] Supervenience is a term that is often used of mental states, where those states depend upon the brain, say, but are more than can be accounted for by those states.

be rational and yet, to the outside world, they may appear to be random in some degree at least. This is because they are based, according to my argument, on internal factors which only the choice-maker knows about. This position allows that an individual decision is rational, in that it is based on reasons private to that individual. However, it does not exclude the possibility that the autonomous core at the centre of the person may be able to make an input to the decision which is independent of all existing mental content. This would not be perceptible to others because it would not be separable from all the other unobservable variables, but it would allow the individual person an individual identity with which free will could be associated. Any such action would have to be either the covert action of God – who is the only other being to whom the choice could be attributed – or a decision of the person made independently of God.

It is the latter that Byl will not allow. For him 'creatures cannot act independently of God' (2003, p. 114, top). This is consistent with his strong view of sovereignty which holds that God is in ultimate and detailed control of everything – including what we say and do. We might, mischievously perhaps, wonder why, if God is truly sovereign, he cannot delegate some freedom to his creatures!

This view of sovereignty is the classical – and extreme – view, the thread of which can be traced through the whole of Christian history; this includes Augustine but is most frequently associated with John Calvin. It is part of a systematic theology which has been the source of profound, and sometimes bitter, debate over the last few centuries. Theologians have wrestled, none convincingly I think, to reconcile real human freedom with God's sovereignty. It has been justly remarked that the volume of literature of the debate through Christian history is entirely out of proportion to the attention given to the subject in scripture – and especially in the Gospels. The good news, that the new life in Christ is open to all, and the example of the Apostles in pursuing that belief cannot be easily reconciled with the, admittedly brief, Pauline pronouncements on predestination. In essence, one has to judge between the literal truth of a few texts (taken out of context?) and the overall thrust of the Gospel. In the present case, the former seems to make nonsense of the latter. The total determinism which Byl wishes us to accept is simply incredible, partly because it ignores the 'level' of creation that we are talking about. In this sphere one quickly runs into logical contradictions. If Byl is correct, the opposition to his view, so widespread in Christianity, must equally be an expression of the will of God – a thought which bears prolonged reflection.

There is one obvious way in which the two views, set in opposition here, might be reconciled. If, as I shall allow later, all apparently chance happenings are really pseudo-chance events, then the sovereignty of God is saved. Everything is ultimately determined by God, yet what happens is no different from what would have happened if chance had been real! In principle, God could predict everything because he knows the 'formula' which generates the random numbers. This is true, but it also remains true that something indistinguishable from pure chance is being used, and pragmatically this is no different from the position I have been advocating. The real point here is that, although God chooses the formula and knows what it is, he does not *need* to know. This means that it is a matter of indifference to him which of the many possible formulae he chooses to use. At this deeper level therefore, there is still a basic uncertainty – and the prospect of an infinite regress if we pursue this line of thinking further.

THE ADVANTAGES OF CHANCE

In the second stage of this chapter I move on from the negative to the positive side of my reaction to Byl's case. It is logically possible that everything in the world is just like those who take Byl's part say it is. The movement of every single particle in the universe could be controlled precisely by God with the whole creation being a wonderfully orchestrated (technically speaking) symphony to the glory of God.[7] If things *have* happened in this particular way, they clearly *could* have happened that way and God *could* have chosen to do it that way. Why then should he have contemplated any other way, much less committed himself to implementing it? Because, I shall argue, it gives greater freedom for God to express his true nature and purpose. In short, it is a more 'Godlike' way to do it.

Let us begin by returning to the creation of planets bearing intelligent life. I pointed out that one could be virtually certain that these would arise 'by chance' if the universe were sufficiently large, and furthermore, that there were likely to be a 'few' planets rather than just one. Is this view of creation theologically credible? At first sight the notion of God's 'casually creating' a few planets here and there might seem an uncharacteristic way of carrying out such a momentous undertaking, but what are the alternatives? Most obviously, and at the other extreme, one might contemplate the fabrication of the present earth in a specially selected corner of the universe by the direct action of the Creator – whatever that might mean. But what then would be the point of the vast universe beyond? Surely not just to impress the inhabitants of earth with the prodigal power of the Creator? Would

---

[7] A useful account of the debate on the question of whether the universe is deterministic is given by Taede Smedes in 'Is our universe deterministic? Some philosophical and theological reflections on an elusive topic' (2003).

it not be more God-like to suppose that a universe on the observed scale was in some way *necessary* for the existence of this planet? If we opt for the 'chance hypothesis', it *was* necessary for the universe to be large enough for humanity to arise with near certainty – but does not this relegate mankind to the edges of creation as a small and insignificant by-product of something much bigger? Not if we can get out of the habit of judging importance by sheer physical size. The vast amount of energy and matter necessary to start the human enterprise is sufficient evidence of the significance of what was being undertaken. Having absorbed all of this, the reader might still be uneasy with the seeming haphazardness of it all. Surely God would care where, exactly, the habitable parts of the universe occurred? Probably not and the reason lies in the hidden assumptions which we so easily transport from our everyday affairs into the uncharted territory of the cosmos. Location may be the prime issue in the domestic housing market but it has no significance in space. The universe looks much the same in whichever direction we view it and from whatever location. So it makes no difference to God, we may suppose, just where these things happen. It is the local conditions that matter, not the cosmic setting. The significance and meaning of a Mozart opera does not depend on whether it is performed in Melbourne or Montreal. It is the character of the creation itself that matters.

These conclusions embody an important principle which underlies much of our thinking about God's action in the world. Roughly stated it says that God does not need to be directly involved in any events whose outcomes have no *relevant* consequences for his wider purposes. This is not an abrogation of his responsibilities but the assumption of his proper place in the scheme of things. Analogies are always incomplete and can be misleading but the development of the analogy of a management hierarchy, which has already been used in chapter 9 (p. 152), may point us in the right direction.

Large and complex organisations such as companies or even nation states may have a single person at the helm – a chief executive or president, let us say. That person has overall responsibility to the shareholders or electorate for whatever happens. Typically their role is to provide a sense of direction and purpose, vision, ideas, judgement and all those things that we sum up in the term 'leadership'. Such a person is not expected to check every invoice or interview candidates for office jobs. Those things are delegated down the line to an appropriate level. The person at the top is directly concerned with the big issues but they are also *responsible* for what happens at all levels and their resignation may be called for if their subordinates fail. They are responsible for creating a system which does the job, not for doing the job themselves.

Something similar may be true of God. He is *responsible* for every single thing but he does not need (nor should he be) involved in executing it down to the last detail. That can safely be left to the autonomous system which he himself has created for that very purpose. We would justly criticise anyone in an organisation who continually interfered with the work of their subordinates. Similarly we might question those who expect God to concern himself, unnecessarily, with those parts of the creation which have been designed to look after themselves.

LIFE ON EARTH

Next we turn again to the arena where the manner of God's involvement has been most hotly disputed, namely the appearance of human life on earth. Is life the natural consequence of the interplay of chance and necessity or was it specially created by God 'with his own hands'? Unfortunately, the issue has become clouded by the way in which the battle lines have been drawn up. This has more to do with questions of how the book of Genesis is to be interpreted than with the deeper question of how, or whether, God is at work. I am not about to rehearse the debate of creationism versus evolution but to look at how God might have acted in the evolutionary process.[8]

I start with three accounts which might be given of evolutionary history. One is that it is the product of a purely deterministic process. On that account any appearance of randomness in the historical record is just that – appearance. Any use we make of probability theory to describe what we observe is just an empirical description of our ignorance, not of any ontological uncertainty. Theistic determinism says that this is the way God intended it to be and his hand is to be seen in everything that happens, however trivial. It covers both what one might call the short-term development that creationism requires and the more long-term view of atheistic determinism.

The second view is the one where most things which happen are actually due to chance, as the modern scientific account suggests, but that God intervenes occasionally to give things a nudge in the desired direction. At particularly significant junctures it may need a particular mutation or other micro event to keep evolution on track. As I have noted several times already, there is no way we can distinguish this model from complete indeterminism because interventions are sufficiently rare to be lost in the greater mass of genuine randomness. Also, as I have noted, it is not clear whether such occasional interventions would be sufficient to do their job. In other words, the method might not work.

---

[8] See note 7 of chapter 11.

The third account is that provided by so-called naturalistic science – where the chance events are what the critics call 'blind chance' as though the adjective qualified the noun in some significant way. It is this third view which creationists and supporters of Intelligent Design have had in their sights as the main threat to theology. It appears to banish God entirely and replace him by the ogre called blind chance. This was never a fully defensible view and it is even less so now than it once was. It was always possible to argue that this was another example of God's acting with the aid of chance. Thus, if life had emerged from the seemingly chaotic goings on in nature then it certainly could do so, and given enough opportunities, it certainly would have done so. The main difficulty arose from the view, championed by Gould, that there were so many paths for evolution to follow that it was incredible that our particular path, the one leading uniquely perhaps to us, should have materialised. It was at this point that many apologists have felt the need to invoke the intervention of God to make sure that the right path was chosen. The ubiquity of convergence, which was discussed in the last chapter, radically alters the need for such interference. For if the paths of evolution converge it becomes much more likely that all (or many, at least) roads will 'lead to Rome'. In addition, if the main sphere of God's activity turns out to be in interacting with human minds, the greater part of evolutionary history may be seen as a preparatory stage which was designed to deliver what was necessary with the minimum of interference from outside. Theologically speaking, this third alternative, of God's creating by chance, may prove to be the most God-like.

This last point brings us to the threshold of a major issue which needs a chapter to itself. Its origins go back to the third example given in the last chapter, concerned with how God might use uncertainty in a constructive way in the realm of human affairs. For those like John Byl who, as I have already noted, believe that there can be no human action apart from God, this question does not arise. But for the rest of us it opens up a whole new vista.

CHAPTER 13

Choice and chance

The discussion of free will and order at the end of chapter 12 brings us up against the central question of whether human choices which appear random can be *really* random. We must, therefore, attempt to unravel the intimate relationship of choice and chance. The first question here concerns in what sense human choices can be said to be random. The idea linking the two is that of unpredictability, which is the key characteristic of both free choice and chance. This leads on to a consideration of how God can 'create' chance and to the observation that total unpredictability, at every level, is impossible.

A PARADOXICAL SITUATION

Choice and chance stand in a paradoxical relationship to one another. If choices are free then, presumably, they are not entirely predictable; for if they were predictable they would be determined, in part at least, and hence not free. If chance implies lack of predictability, then chance events could hardly be the result of deliberate choice. Yet rational choices, when viewed collectively, do often appear as if they were random. If all happenings, which to us appear to be completely random, were to be attributed to God, then they would have to be his deliberate choices, rationally made, so it is difficult to describe them as totally random.

The purpose of this chapter is to unravel the confusions which lie hidden in such assertions. These issues have surfaced several times in earlier chapters but the time has now come to make them the focus of our attention. This will involve going over some old ground again but will serve to emphasise its importance in the present context. I begin with the matter of human choice, leaving God out of the picture for the moment.

ARE HUMAN CHOICES REALLY RANDOM?

This appears to be a very silly question. It is of the essence of our human nature at its best that we make informed and rational choices; how then can we speak of them as random? To some scientists, however, there is nothing silly about the idea. Marvin Minsky (1987, p. 306),[1] for example, has put the matter categorically.

We each believe that we possess an Ego, Self or Final Center of Control, from which we choose what we shall do at every fork in the road of time. To be sure we sometimes have the sense of being dragged along despite ourselves, by internal processes which, though they come from within our minds, nevertheless seem to work against our wishes. But on the whole we still feel that we can choose what we shall do.

He goes on,

According to the modern scientific view, there is simply no room at all for 'freedom of the human will'. Everything that happens in our universe is either completely determined by what's already happened in the past or else depends, in part, on random chance. Everything, including that which happens in our brains, depends on these and only these:

A set of fixed, deterministic laws. A purely random set of accidents.

---

[1] Minsky's remark was made twenty years ago towards the end of a distinguished career. It may be that his view would not be widely held today, or be expressed in such categorical terms, but the quotation serves to define an extreme position.

There can be no room on either side for any third alternative. Whatever actions we may 'choose', they cannot make the slightest change
in what might otherwise have been *because those rigid, natural laws
already caused the states of mind that caused us to decide that way.* And
if that choice was made by chance – it still leaves nothing for us to
decide.

To digress for a moment, any statement like that raises
profound questions about the meaning which can be attached
to it. For if Minsky is speaking of all minds, then presumably
his own is included. If what comes out of his mind is solely
the result of deterministic laws and random chance, it is not
clear why such an automaton's utterances should correspond
to anything in the real world or, for that matter, why similar
automatons like ourselves should believe them. The same
problem arises whenever a human mind attempts to make
statements about the class of human minds to which it, itself,
belongs. This chapter is no exception and rational discourse
can only continue if we reject Minsky's view. It is not open
to us to set ourselves above the debate by supposing that we
alone have a privileged position which exempts us from the
necessity of applying our conclusions to ourselves!

That being said, it is an empirical fact, already noted, that
human choices, when viewed collectively, are often virtually
indistinguishable from random choices.[2] How then do we
reconcile this collective character with what we believe to be
the purposefulness of our own individual choices?

---

[2] It must be emphasised that randomness is not a property of a single event
but of a collection. To say that human choices appear to be random is,
therefore, a statement about a set of actions and not about any single act.
To justify such a statement we have to observe a collection of choices made
under similar conditions and to find that the relevant observed frequency
distributions were what they would have been if they had been generated
by a known random process.

The clue to the resolution of the paradox lies in the way that I have defined chance: the essence of a chance or random event, remember, is that it cannot be predicted, even partially, from anything we can observe about the state of the world past or present. In the case of human choice, this would certainly be true if everything a person thought or did was, as Minsky supposed, the result of random firings in the brain (any partial dependence on any other fixed aspect of the world would reduce, but not eliminate, the uncertainty so does not affect the point being made). The obvious way to reconcile choice and chance is thus to go along with Minsky, and those who think like him, and to say that they are the same thing.

But there is another way to do it which does not rule out at the start the notion of a rational-thinking human being who makes real choices. Suppose instead that the individual's choice is affected by factors not detectable by external observers but known only to the subject. In that case the choice making could be perfectly rational yet the rationale would be hidden from anyone external to the individual. To clarify this statement I refer back to the distinction between a person's external world and their internal world which I made in chapter 12. The internal world may be partly constructed from what happens in the outside world but will also have an input from the 'self'. The individual's choices, as observed from outside, are therefore partly unpredictable and, in the aggregate, will display the same characteristics as a process in which there is a truly random element.

In practice, of course, the inner and outer worlds will become mixed up because of the two-way traffic between them. Something of the internal world will be revealed in patterns in past choices, and there will be a modification of the inner world as information from outside is absorbed to become part of the inner world. It may also happen that choices appear random to us only because we have not noticed causal factors which are operating. Indeed, one may suppose that this is normally the case since our knowledge is so limited.

None of this precludes factors in the external world from having a causal role in what we decide. Genetic and environmental factors immediately come to mind. Some people go as far as to say that they are wholly responsible[3] but, if this were so, one might expect our choices to be much more predictable than they actually are. But this fact certainly does not rule out the possibility that there may still be a large component of unpredictability in our behaviour, arising from the factors which are exclusive to ourselves.

We are very familiar with what we regard as the mixture of determinism and free will in much of what we do. Much market research, for example, depends on the fact that there are things which we can observe about people – their income, hobbies, and so on – which are useful predictors of purchasing behaviour. Voting behaviour is another example of something which is partially predictable. Income, occupation and place of residence all tell us something about the party for which someone will vote – especially at national elections. These predictors may have been better in the past than they are today and their effect will vary from one culture to another. But, when all of this has been taken into account, all voters, for example,

[3] It is common for people to suppose that human decisions are explained by a mixture of 'nature' and 'nurture'. Most of the arguments which take place are about the proportions of each in the mix. If this were literally true, all such decisions would, in principle, be predictable. Once the genetic and environmental factors were identified and exhaustively measured, the decision would be a foregone conclusion. One would have expected there to be a reasonable number of non-trivial examples where outcomes were well predicted but such examples are not easy to find.

believe that they have a real choice and that this is not totally constrained by their economic and social circumstances. In short, they will think of voting as an exercise of their free will. Free will, in the libertarian[4] sense, at least, requires that we should be able to make choices that are not wholly determined by factors which are, in principle, observable by others.

All that I have said so far can be accommodated within a naturalistic framework. No appeal need be made to God to explain the unpredictable choices which people make. The essential point thus far is that the individual has a world of private experience which permits rational decision making and which is not accessible to outside observers.

However, this leaves out of account two matters which are important if we wish to bring theological factors into the discussion. The first adds another dimension to the naturalistic interpretation already discussed above. If God acts through individuals, and if his mind is hidden to us, the unpredictability we observe in the actions will be, partly at least, a reflection of God's participation in those actions. The second case is where apparently random things happen without any human participation at all as, for example, when particles are emitted from a radioactive atom. This takes us back to God's action in the physical world and it may seem rather remote from the sphere of human action. Nevertheless it is convenient to see it as a limiting case of the problem of human decision making which arises in the case when we eliminate the human element and thus leave God as the only remaining decision maker. Can God also do unpredictable things – unpredictable, even by himself? I take the two cases in the order in which I have introduced them.

---

[4] The philosophical doctrine that human choices are neither causally determined nor random. They are genuinely free.

How God acts in the world is a major problem by any standards and this problem was discussed in chapter 9. The most natural way to accommodate God's action in the scheme of things is to suppose that he acts partly, if not exclusively, through interaction with the human mind. According to this view, God's acts would appear under the guise of human acts and thus be indistinguishable from them, in principle, at least. John Byl, and those who think like him, believe that no one can act apart from God and hence that there are no genuinely free choices. According to them our apparent freedom arises from the fact that our choices are those which God has made for us, and which are compatible with our wishes – for which he is also responsible. Even if we take this view, it remains true that the choices which God makes in us will not be predictable and hence, in the aggregate they will appear as if they had been made at random.

At first sight the situation in the physical world is much simpler because the entities with which it deals do not have an 'internal' world. The fact that they appear random is simply a reflection of the fact that we know of no other observables in the world which would have any predictive value. However, on closer inspection, the matter is not so straightforward. From the theological perspective there is no other *person* involved, so any decisions made have to have been made by God. This brings us back to the question of how God might determine the outcomes of physical processes, bearing in mind the fact that, because there is no separate physical cause, they will be totally unpredictable and hence apparently random. How then can God purposefully behave as if he had no purpose? I can link this question to another which arises out of my central claim that God uses chance for his own purposes much as we use it for ours. In fact, these are really the same question which may be posed as: *how can God create chance?*

HOW CAN GOD CREATE CHANCE?

We now turn to look at things from God's side. If there is chance in the world and if God is somehow responsible for it then what happens is, in a certain sense, his choice. How can what he *chooses* be random? To clarify this conundrum we have to think again about how chance events come about.

At this stage we are thinking primarily of the natural world, so I use again the familiar example of the emission of particles from a radioactive substance. Here we are back to the quantum world where, to the best of anyone's knowledge, there are no predictors available. Thus, if we suppose that emissions occur *at random* we know that we can deduce the form of several observable distributions, as reported when discussing the same situation in chapter 3. For example, if we measure the waiting time to the next emission it turns out that the proportion of occasions on which we shall have to wait more than a minute, say, is given by a simple exponential expression. Alternatively, if we count the number of emissions per minute, the frequency distribution will follow Poisson's law. There is an important feature of random processes which has been noted above and needs to be reiterated here. This is, that when we observe randomness, there is independence between events. This does not need to be explicitly mentioned because it is covered by our requiring that the past history have no predictive value. For example, the time of emission of a particle does not depend in any way on the timing of other preceding emissions, and that is what is meant by saying that it is independent of them.

If God determines outcomes then his actions are part of what physics studies. If pure chance is part of the created order and is therefore a work of God, how does he do it? Put another way, if God is the cause of everything, how can he cause something which, by definition, appears to have no cause? This problem was identified in *God of Chance* (Bartholomew 1984, p. 102) and commented on by Byl (in Byl 2003, p. 107) and others, but its importance does not seem to have been generally noticed. If, for example, quantum happenings have all the characteristics of chance events with no discernible cause, how does God, I repeat, bring them about?

It is always possible, of course, that we are deluding ourselves and that there are no purely chance events. We can produce pseudo-chance events, and if we can do so, then they would surely not be beyond God. But let us suppose, for a moment, that there are events which happen independently of the state of the universe. (In passing, it is worth noting that there is also a problem here for non-theists, but for them it is part of the broader question of why there is anything at all. This then becomes the point at which they come to an *impasse* and have to admit that rational enquiry can take them no further. Theists, on the other hand, have the no less difficult problem of explaining why, in God's world, some things can happen without a cause.)

We should make considerable progress on this question if we could see how God might engineer randomness in something as simple as a sequence of random tosses of a coin. Coin tossing is not random in the purest sense but is a very subtle example of a pseudo-random series. Tossing a coin is a dynamic matter where the impulse imparted to the coin, the air currents and a multitude of other physical factors determine the outcome according to the laws of matter which are well understood. Nevertheless, coin tossing is near enough random to provide a convenient and familiar framework in which to discuss the problem. The question now becomes: how might God generate a sequence of heads or tails (or zeros and ones) which was purely random?

The following very tentative suggestion is based on the idea that if causal forces must be present yet have no effect, their individual effects must, in some sense, cancel out. We might think of it as a kind of tug-of-war in which the two physical propensities to produce heads or tails might just balance. Since there *has* to be one outcome or the other, and since there is nothing to decide between the two options, what actually occurs is a matter of chance. Buridan's ass[5] provides a familiar picture of the situation. Placed between two equally tempting bales of hay, where each is equally accessible, the ass has no reason to prefer one over the other. Which way the scales are tipped cannot then be predicted from a knowledge of the two forces involved moving in either direction because they are equal but opposite.

A much simpler solution to the dilemma, touched on above, is to suppose that God, like us, can produce pseudo-random events, not genuinely random events. That is, behind every happening for which we can see no causal factors, there lies, deeply buried in the mind of God, a pseudo-random-number generator. For those who demand that God be in total control, this provides the determinism they are looking for. But in interpreting this we must recall what I have repeatedly said about 'levels'. What is random at one level may be determined at the next higher level. Chance and determinism are not mutually incompatible alternatives but each may imply the other – but at a different level. This is tantamount to saying that total randomness is impossible anyhow.

---

[5] The dilemma of the ass which starves because of its inability to make a decision often comes into philosophical discussion. It dates from the fourteenth century but is said to have been discussed by Aristotle (384–322 BC).

TOTAL UNPREDICTABILITY AT ALL
LEVELS IS IMPOSSIBLE

I have repeatedly stated that lawfulness in the aggregate may be the product of randomness at the individual level. What is unpredictable at the latter level, therefore, necessarily becomes predictable at the higher level. The converse may also be true but we are not at liberty to have unpredictability at both levels simultaneously. This fact places restrictions on a God who operates at the individual level, and thus places a question mark against the total freedom which we might require to be possessed by someone who was truly God.

To make the point clearer, let us return yet again to the emission of particles from a radioactive source. Nothing we can observe about the process – such as the length of time from the last emission – has any predictive value for the timing of the next emission. One consequence of this is that the frequency distribution of the intervals between emissions has a characteristic and predictable form known as the *negative exponential*. It looks like the example given in figure 3.1 (see p. 35), which shows, among other things, that the frequency declines as the length of interval increases.

Next, suppose that the actual timing of the emissions is determined by God. In order for his choices to conform to the empirical facts, it must not be possible for the human observer to detect any departure from randomness. Detection would be possible if the emerging frequency distribution departed from the negative exponential form. We can therefore imagine the Deity having to keep an eye on the overall frequency distribution at the higher level, as well as making sure that nothing in the detailed history of the process at the lower level, including the recent inter-event intervals, reveals any non-independence. It would surely be much simpler and just as effective for the Deity to buy or borrow a pseudo-random-number generator and let that determine when emissions should occur! In other words, it is not possible to do better than a randomising device, so why should God not use the best tool available? What would it tell us about his nature if he chose the second best?

For all things that depend only on the aggregate behaviour of random events there is no obvious advantage in actually determining when each individual emission occurs. Where individual happenings *might* matter is in circumstances where they trigger some macro event, perhaps through the amplifying effect of a chaos model. A tiny proportion of deliberately engineered events would then pass unnoticed in the mass of randomness and those could be inserted by the Deity at will, with no possibility of detection. This puts God's particular actions beyond the reach of science, and it promises to disengage the two fields from potential conflict. However, this raises the awkward question which I posed in connection with the coin-tossing example in chapter 8. This whole manner of working smacks of subterfuge, which is hardly a God-like characteristic. But maybe there is no other way, given the other requirements of running the whole creation. Further, it is not clear why the transparency, so often demanded in human decision making, should necessarily be part of the divine character. It does leave open, however, the question of why such fine-tuning should be necessary except, perhaps, to counter the unwanted consequences of human free will. But then we must ask why a more direct intervention at the level of human decision making might not be more effective.

CHAPTER 14

God and risk

The central thesis of this book is that God uses chance. This appears to carry the implication that God takes risks. This simple statement has repercussions for most parts of theology but here we focus on the central issue. What is required is a theology of risk and this is what the chapter aims to provide. After some theological preliminaries, I commend a view which it is proposed to call critical orthodoxy.

DOES GOD TAKE RISKS?

We live in a society obsessed with risk. Risk assessment and risk management are part of everyday life in business and industry. We are exposed to all manner of hazards, not only to life and limb, but to our comfort and general welfare. The insurance industry has long existed to alleviate the problem by spreading risk but the threat of litigation and crippling damages has heightened awareness of the problem. It is hardly surprising that there should be a ready market for Peter Bernstein's book *Against the Gods: the Remarkable Story of Risk* (1998).[1] But risk also raises questions for theology. Gregerson (2003b) has taken steps towards a theology of risk and followed this up (Gregerson 2006) with a study which, among

---

[1] Bernstein's book also contains much else that would provide useful background reading for this book, about sampling for example.

other things, challenges the views of some contemporary sociologists.

All of this work looks at risk from the human side. There is also a God-ward side on which we concentrate in this chapter. The idea that God uses chance poses many problems for theology, and the chief of these, perhaps, is the question mark it places against the sovereignty of God. This challenge has occupied us a great deal in the earlier part of the book, especially in chapter 12. It also lies behind many of the issues discussed in this chapter but I shall keep it in the background as we approach things from a different angle.

If my thesis that God *uses* chance is accepted, we cannot avoid the consequence that he appears to be taking risks. However, it may not be immediately obvious why this is such a problem. After all, the whole thrust of the argument has been that chance can be used to achieve determinate ends, so we might naturally wonder what risks are being run. To clarify this we need to recall the differing levels of nature and society. Near certainty at one level may be the product of uncertainty lower down, and it is at that lower level that outcomes may have unwanted effects. If it is true that the evolutionary progression to ourselves is the product of chance and necessity, then it is painfully obvious that this is not achieved without waste and suffering along the way. If that is so, God could not avoid taking risks. To get the global picture right, it appears that it may have been necessary to get many of the details wrong. This is our starting point.

Two questions immediately pose themselves. First, is God really a risk taker?, and secondly, why did he create a world in which we are exposed to so many risks? The aim of this chapter is to suggest answers to both questions but first we need to be clearer about what they mean and why they matter.

To say that God is a risk taker means that he takes – or has taken – actions where the outcome was intrinsically uncertain and which might turn out contrary to his intentions. It is immediately clear why this is a serious question for Christians. A God, in the classical mould, who is omnipotent and omniscient, ought not to be at the mercy of his own creation like this. Such a God ought not to need to take risks. Even if, for a moment, we concede that things might occasionally get out of hand, God should surely have the power to get things back on track before any harm is done. Indeed, the language of risk seems to have no place in describing the nature and role of the God of the Bible and Christian orthodoxy.

Such a God does not fare much better if we judge him in relation to the second question. Many of the risks we run lead to immense suffering and damage. Even when we allow for the fact that much of this is self-inflicted, the residue is still capable of supporting a serious charge. Either God deliberately created a world in which there was bound to be much hurt and suffering, or he did not have sufficient control over things to stop them going awry. In the former case we have to question his very nature, and in particular, the Christian claim that his nature is love. In the latter it is his omnipotence that is undermined. Furthermore, if he cannot fully control things, it is difficult to see how he could act effectively in the world and so his providence has to be added to the list of orthodox casualties.

If we take all this into account, the credibility of the whole edifice of Christian orthodoxy begins to look decidedly shaky and there is no lack of critics on hand to drive that fact home to the Christian's disadvantage. But perhaps there is no cause for alarm after all, for these gloomy prognostications only follow if we give a positive answer to the first question. *If* God is a risk taker then we may indeed be in trouble, but surely centuries of Christian theology cannot be so lightly over-turned? Should we not take our stand on the certain rock of revelation and dismiss all of this as idle speculation? I think not. And my reasons lie in the fact, which has repeatedly been brought to our attention, that there is much in the scientific picture of the world that seems to be inconsistent with a theology which sees God in detailed charge of every single thing. I have touched upon these matters in earlier chapters but now I tackle them in a more systematic way. Although they are usually dealt with in the context of God's sovereignty, I wish to discuss them from first principles under what I shall call *naive orthodoxy*. If we take the new knowledge about the role of chance in the world seriously, we must either revise our view of God or join those who have abandoned him altogether.

Earlier generations would have been puzzled by such a claim because, from their perspective, the boot would have been on the other foot. For them science seemed to involve replacing uncertainty with lawfulness, which was readily accommodated to the Bible's view of an omnipotent God. At the very core of science was the lawfulness of nature typified by Newton's laws of motion. And have not Christians, including Newton himself, seen this as testifying to the faithfulness and reliability of the God who created it all? In the almost mechanical necessity of the physical world, there seemed to be no place for uncertainty and risk taking. To be sure, this view was essentially deism and posed other problems for the believer, and some ingenuity was needed to see how room could be found for special providences. But the general providence expressed in the laws of nature seemed secure and no one could level a charge of capriciousness at the God to whom science seemed to point.

Times have changed and science has moved on. It is not that the laws of science have turned out to be wrong. They are as valid as ever they were. But we now see them as a partial description of an immensely complicated universe in which chance plays as important a part as necessity. Before we can begin to reconstruct a credible theology, which is consistent with our new knowledge, we must recall the different kinds of uncertainty which we encounter in the world. I have identified three broad categories: the pure chance which we observe primarily at the quantum level; accidents which include the mutations of evolution; and, finally, the uncertainties of human decision making.

SOME THEOLOGICAL IMPLICATIONS OF THE WAY THE WORLD IS

Given that the world is such an uncertain place, we face some serious questions about the apparently hit-and-miss nature of some aspects of the creative process. Leaving aside the question of the hazards attending the appearance of a habitable planet, which I have already dealt with, let us focus on the emergence of life. According to the likes of Stephen J. Gould, Jacques Monod,[2] Richard Dawkins and others, the appearance of life was a chance happening, depending on accidental copying errors in the reproductive process. If that was all there was to it, Gould's metaphor of the rerun film applies and our presence is the almost incredible result of a random process. On this reckoning, it was extremely unlikely that intelligent human life would have been the outcome. That hardly sounds like the strategy of a loving God intent on creating beings destined for fellowship with himself for, if it was, it was almost

---

[2] Jacques Monod's book (1972 [1970]) is, perhaps, the original and clearest exposition of the thesis that chance and necessity are both necessary and sufficient to account for all living things. Dawkins and Gould are but two of the most widely read authors who have propagated the idea.

certain to fail. However, convergence comes to the rescue by showing that the appearance of sentient beings, not so very different from us, may have been almost certain.

Even if we take the view that the emergence of life on earth was inevitable, there were still enormous risks to be run. Collisions of the earth with meteors or asteroids could have destroyed the fledgling life or so distorted the path of evolution as to close the door to human development (something similar seems to have happened to the dinosaurs). Volcanic eruptions, radical climate changes or devastating diseases could likewise have posed insuperable barriers for the progress of life towards its intended culmination in human beings.[3]

Again, even if, against all the odds, human life did become firmly established, the hazards of human history seem to put the whole enterprise at risk. We know from that small segment of society with which we are intimately acquainted, how much often seems to depend on the merest whim. When magnified a millionfold onto a world scale with its recessions, wars and conspiracies, it seems incredible that anyone, even God, could control what was going on. On the face of it, at least, human history does not look like a carefully orchestrated divine drama moving inexorably to its intended end. The threat of global disaster by nuclear war, famine or epidemic is still sufficiently real to give us serious pause for thought.

The dilemma for the theologian is most acute when we come to consider the place of the Incarnation in the scheme of things. Think of all the mishaps which can occur in childhood to nip a promising life in the bud. If, as is claimed, only 50 per cent of children at the time of Jesus reached their tenth

---

[3] The possibility that a burst of gamma rays might be responsible for a mass extinction 443 million years ago was raised by Mark Buchanan in *New Scientist*, 30 July 2005.

birthday, the risk of pinning the redemption of the world on a single individual is apparent. Did God have to take these risks? If he did not and Jesus was spared those risks, how can we say that Jesus was truly human? For experiencing risk and being changed by the experience is part of what it means to be human. The doctrine of the full humanity of Jesus seems to preclude the kind of special protection accorded to royal personages and their likes today.

These considerations, and others like them, seem to take us into a very different world from the one addressed by the prophets and New Testament writers of the Bible. It is not easy for those immersed in contemporary culture, in which risk seems so deeply embedded, to take seriously the Christian story as naïve orthodoxy presents it.

BENEFITS OF A RISKY WORLD

We have assumed, perhaps too readily, that risk is a bad thing and that its ubiquity is an embarrassment to the believer. This is far from the case. In evolution, for example, the almost prodigal variety which nature throws up seems a terrible waste from an economical human perspective, but it confers many benefits. It provides an insurance against the unexpected. For example, by ensuring that there is a wide range of variation among organisms of a particular type, it increases the chance that there will be at least some survivors of any disaster capable of coping with the new situation. This use of chance to counteract the effects of chance is a subtle and surprising feature of nature that has profound implications.

When we move on to human society there opens up a whole new dimension of risk. Risk is not only something which we seek to avoid or to insure ourselves against. It is also something that we seek out, or even create, for ourselves.

Mountaineering, caving or extreme physical sports are not thrust upon us, and though they may cause death or serious injury, there is no lack of people eager to take part, and prizes are awarded to the successful. Likewise, no rational person guided by economic criteria alone would gamble, because the expected gain is never positive. Nevertheless, to many the excitement and suspense when much is at stake seems to make it worth while. The enduring appeal of many games depends on just the right balance of skill and chance. In snooker, for example, there would be little interest at the highest level if chance played no part, and it is a prime function of rule makers to create just that right balance. The rules must favour the skilful but involve enough luck to leave the outcome in doubt and keep interest alive.

Although games might be considered a rather trivial matter to bring into this discussion, we have seen already how illuminating they can be. Games mirror life, and in some cases, perhaps, provide a substitute for the thrills necessary to our wellbeing of which we have been deprived by modern civilisation. The competitive instinct is deeply rooted and seems to be essential for progress. It could be plausibly argued that risk is a necessary ingredient for full human development. It provides the richness and diversity of experience necessary to develop our skills and personalities. This does not mean that the risks are always welcome at the time. We can all look back and identify occasions of great uncertainty which we would have gladly avoided but which are seen, in retrospect, to have contributed to our development. The notion of trial and test in the religious sense, much of which involves facing hazards of various kinds, is familiar enough to Christians. The story of Job springs to mind, as does the temptation of Jesus and his experiences leading to the Cross. The remarks in 1 Peter 1 about faith, like gold, being tested by fire, must have found many echoes in the experience of the early Church. It is not easy to imagine a world without risk or to know what its absence would imply. At the very least we should now be aware that its abolition might not be wholly beneficial.

THEOLOGICAL PRELIMINARIES

I indicated at the beginning of the chapter that chance and uncertainty were widely seen as a threat to what I called naïve orthodoxy. The question now to be considered is whether we can begin to fashion a more critical orthodoxy that comes to terms with the new without abandoning what is essential in the old. I want to suggest that we should welcome the new because its insights are almost wholly beneficial.

As a preliminary step it may help to make some remarks about the nature of belief. First we must reject the *list* approach, according to which orthodoxy consists in assent to a long list of propositions. From this perspective the 'true' faith is then that which is held by those who assent to all items on the list. Those of weaker faith place crosses or question marks against some items and are seen as professing a somewhat watered down version of the faith; sects are distinguished by the items on the list to which they attach particular importance.

This is, of course, a caricature but something like it lies behind much of the posturing in religious circles. It follows naturally from what we might term the mathematical or deductive approach to knowledge.[4] According to this, certain truths are held to be given, such as the self-evident axioms of geometry. Theology, like geometry, then consists

---

[4] Foundationalism is a term often used to describe the equivalent way of reasoning in theology. One starts from certain propositions which are taken as 'given', in some sense, and which therefore provide the foundation for the theology that is built on them.

in working out their logical consequences. Propositions can be established as definitely true or false by the processes of logic, and the aforementioned list then consists of those truths which can be deduced from the axioms provided by revelation. We do not need reminding that there is some diversity of view about where the axioms are to be found and what they say.

To my mind, the process of scientific inference provides a better model for theology. Not so much in the process of conjecture and refutation advocated by Popper[5] – though that has its merits – but in the inductive accumulation of knowledge. I have advocated and illustrated this approach in my book *Uncertain Belief* (Bartholomew 1996), arguing that all knowledge is uncertain, in varying degrees.[6] This is a bottom-up approach. It yields nothing to its rivals in the importance it attaches to the treasures of Bible and Church. The 'faith once delivered to the saints' is expressed in the record of writings, experience and practice of the Christian community. This is an essential part of the evidence with which the theologian must grapple in trying to construct a coherent account of reality but it is not the whole of the evidence. It must be *interpreted*

---

[5] Karl Popper started from the idea that, though one could never determine with certainty what was true, it was sometimes possible to be certain about what was not true. Science should thus proceed by a process of *falsification*. For example, one could never be certain that all crows were black but the proposition would be falsified if a single crow were to be observed which was not black.

[6] Richard Swinburne pioneered the inductive approach in theology using Bayes' theorem. The first edition of his book *The Existence of God*, published in 1979, was a landmark publication in this field, though, for reasons set out in Bartholomew (1996), I did not think the application of the method was wholly convincing. The second edition, published in 2004, was described by the publisher as 'the definitive version for posterity'. In the preface the author acknowledges that his 'critics are 'many' and that they 'have provided much help' but it is not clear what form that help took.

by reference to the knowledge from all sources including, especially, science.

Pictures may help to make the distinction between the deductive and inductive approaches to belief clearer. The approach of deductive theology is like the house built on the solid rock of certain truth. As long as the foundation remains sound, the elaborate edifice above stands firm and secure. But if the integrity of the foundation is compromised, cracks begin to appear and although appearances can be maintained for a while by papering them over, the pretence cannot be kept up indefinitely.

What about the inductive approach? Is that like the house built on sand, as it will seem to many? In a way it is, but a better analogy is the picture of a railway built across a bog which was used on the cover of *Uncertain Belief*. When George Stephenson built the Manchester and Liverpool Railway he had to cross Chat Moss. The obvious way of building the line would have been to sink piles onto solid rock beneath and support the line in that way – but no such foundation could be found. So he floated the line across on a bed of heather lashed to hurdles. The weight was thereby distributed so widely that even such a frail foundation as boggy ground could support it. The strength, in this case, was derived from the mutual support of the interlocking hurdles, none of which was excessively loaded. I think it is better to regard the basis of belief in these terms. Although no single element will bear much weight if looked at in isolation – as many critics have shown only too well – the combined strength is immense. It is the cumulative and interdependent effect of all the fragments of evidence which ultimately provides the securest foundation.

One important consequence is that the near certainties only begin to emerge when we stand back and look at the whole picture. In many of the details we shall be mistaken without always being able to tell which. It is in the common strands that run through time and across diverse cultures that the essential core is to be discerned. To take his words only slightly out of context, John Wesley put it rather well when in his sermon on the catholic spirit, he said:

No man can be assured that all his opinions taken together are true.
　　Nay every thinking man knows that they are not . . . He knows, in the general, that he himself is mistaken; Although in what particulars he mistakes, he does not, perhaps he cannot know.

We can take the analogy with scientific method one step further. It suggests how we may expect doctrine to develop as knowledge advances. Science typically progresses not by replacing old knowledge by new, but rather the old finds its place in the new as a special case or an approximation. Newton's laws of motion are no less true today than in the seventeenth century. They are implicit within relativity theory as an approximation valid in the normal range of human perception. Although quantum theory was designed to deal with the world of the very small, it is consistent with the mechanics of the human-sized world. The new in each case includes the old but has a wider range of validity, encompassing the very large, the very small or the very fast. Similarly in theology, the old orthodoxy will remain, but will now be seen as part of a fuller truth valid over a wider range of time and culture.

We have seen this happen already with the doctrine of creation. Originally formulated in terms of the small world of the Mediterranean basin and a primitive cosmology, it has developed as our scientific horizons have extended. It is now an altogether greater thing and the God to whom it points is thereby magnified. The God whose space is measured in light years is immeasurably more magnificent than the tribal deity of the early parts of the Old Testament.

Much the same could be said of evolution, though here there are added complications arising from the resurgence of creationism and the advent of the Intelligent Design movement, whose claims were examined in chapter 7. The process of understanding here is still growing and subject to lively debate, especially in the United States of America.

Often new developments in human thought, such as the evolution of life, have been seen as threatening to belief because of what they appear to deny. Only as we become accustomed to them and fully absorb their implications do we begin to see that our understanding of God and his world has been enlarged. The doctrine is not diminished but greatly enriched. The new orthodoxy is richer, not poorer, as a result.

THE THEOLOGY OF RISK

Can the threat to God's sovereignty, apparently posed by risk taking in all spheres, be treated in the same way as in creation and evolution? Can a world in which chance seems so threatening to our understanding of God really turn out to be friendly to a more critical orthodoxy? If it turns out that we can claim that the world of chance and theology are compatible, can we go further and claim that our understanding of God has been deepened and enriched by the new knowledge? In the remainder of this chapter I hope to show, in outline at least, that we can.

So we return to my opening question: 'Is God a risk taker?' In the light of my discussion this question has to be worded more carefully because we have to distinguish between ultimate goals and short-term deviations. We have seen that determinate ends may be achieved as the result of averaging many random effects or by the interactions within the process. This means that the end of the process may be virtually certain, even though the path to that end is not determined. Is it sufficient to preserve our understanding of God's greatness that he, as it were, gets there in the end – or that he must never put a foot wrong? It seems to me that it is the end that matters and if deviations towards that end deliver side benefits, the net result may be gain.

To begin with creation, suppose someone claims – as many have – that the probability of a life-bearing planet on which something like ourselves would emerge is so small that no God could have used such a risky procedure to create it. According to my earlier analysis, this is simply not true. Maybe God's purposes did not require it to be on this particular planet at this particular epoch that we made our appearance. Perhaps there are innumerable other places where it could have happened. Provided only that the number of possible times and locations was sufficiently large, the chance of ultimate success could, as we have seen, be made very close to certainty.

If we move on to the evolution of life on earth we have seen that this, too, may not have been such a risky undertaking as Overman and his kind have supposed. There is the possibility that life might have been almost certain to arise in the conditions of the primeval earth. Apart from the chemical evidence there is Kauffman's work on self-replication, which suggests, even if it does not yet prove, that autocatalytic processes are capable of producing self-replicating entities. But once life got under way for whatever reason, the phenomenon of convergence seems likely to have severely limited the number of possible outcomes, making sentient beings like us almost inevitable. God is not bound by our notions of economy and it enhances rather than diminishes our place in the scheme of things to know that it took a universe as big as this to bring us into being. The risk of ultimate failure which God took would then be negligible. As Stuart Kauffman puts it, we are 'at home in the universe'.[7]

But if the merest hint of a residual risk leaves us feeling uncomfortable, perhaps we should reflect that if God is greater than anything we can conceive, it might be the case that any enterprise worthy of his nature would have to push at the very limits of what is possible.

More worrying, perhaps, are the historical risks, most notably of the Incarnation. If the redemption of humankind required the ultimate conflict between good and evil expressed in the Cross and Resurrection, and if this could not be guaranteed under normal conditions of human existence, where does that leave us? The position is only slightly eased if we allow that the final conflict might have resolved itself in a variety of equally effective ways. It raises the unanswerable question of whether there could have been as many attempts as necessary. There seems to me to be no overwhelming theological objection to such an idea but we should not take this option too eagerly as there are other alternatives. There is the important principle of likelihood inference (or inference to the best explanation, in the language of philosophers) which I have used before and which comes into play again. I have also dealt with it at some length in two earlier books (Bartholomew 1984 and 1996) so will not go into details now but only note that it has wider applications. If we had made reasonable calculations *prior* to the

---

[7] This is the title of Kauffman (1995) which was the 'popular' version of Kauffman (1993). The title was designed to convey the core idea of the earlier book that life may have been an almost inevitable outcome of the complexity of the primeval world. Self-organisation was the key principle at work.

birth of Jesus we would doubtless have concluded that the Incarnation was an extremely risky event. But we do not stand at that point.[8] We have to consider the question with all the information available to us now — and that includes our own existence. That fact has to be brought into the argument. The likelihood principle used says that we should judge contending hypotheses by the probabilities they assign to the actual outcomes, not those which might have occurred but did not. Therefore any hypothesis which gives a relatively high probability to the occurrence of a 'successful' incarnation is to be preferred to one which does not. Those probabilities which, calculated in advance of the event and much favoured by scientists untutored in statistical inference, give extremely small probabilities to life and historical happenings should therefore be rejected in favour of any hypotheses which make them more likely. The relevance of this argument to the risks of the Incarnation is that since all the things which might have thrown it off course did not, in fact, happen then we should favour hypotheses which do not make it appear so risky. I do not pretend that this is easy, because of the need to affirm the full humanity of Christ, but it should cause us to exercise caution in rushing to conclusions. Any suggestion that we might go beyond likelihood inference and use a Bayesian analysis runs into the insuperable problem of enumerating the contending hypotheses and assigning prior probabilities to them.

---

[8] One way of characterising the difference between the likelihood approach and the frequentist approach of Neyman and Pearson is by reference to the stage in the process at which probabilities are calculated. If we wait until the data are available we consider the probabilities which a range of hypotheses would assign to those data. In the frequentist approach we calculate the probabilities of the different outcomes which are possible before they happen.

The next risk to consider is the one which God took in creating a world where there would be autonomous beings capable of exercising their own will and acting contrary to their maker's intentions. In this case the very notion of free will implies a risk, but again we have to distinguish the local and individual effect from the global and collective. The Law of Large Numbers may, again, determine long-term outcomes, not in spite of, but because of the exercise of individual freedom. When to this we add the positive gains for human development, the flexibility and adaptability that require the world to be a risky place, the case may have been overwhelming for doing it this way, even supposing that there was any choice in the matter.

But does the world need to be such a risky place given the immense amount of suffering which this seems to entail? In responding it is customary to distinguish 'natural' suffering from that which can be laid at the door of humankind. We are only now beginning to recognise the long-term and far-reaching damage that can be done by the human race. Nowadays we not only have to put the harm that one does to another on the human side of the ledger but increasingly it is becoming clear that the many so-called natural disasters, involving climate and ecology, are the (often unwitting) consequences of human greed and ambition. Nevertheless much remains in the realm of accident which cannot obviously be blamed on anyone. However, having abandoned the deterministic world-view, we cannot lay the blame directly on God either – a view for which there is good scriptural warrant (in Luke 13. 2–4; John 9. 3, for example). This is a considerable help because it provides us with an answer to the oft-posed question: 'Why me?' The paradoxical answer is 'For no reason whatsoever'; the suffering is not a targeted response on the part of God to some specific misdemeanour. This is a very positive benefit for theology. We are familiar with the moral difference between having direct responsibility for some heinous act and a more general responsibility. For example, we do not accuse the Prime Minister of being personally responsible for every personal misfortune which results from government legislation. He and his government are, of course, responsible in a general sense, as God is, but the moral implications are not the same.

So we are left with the conundrum with which we started. We have to explain why there is so much suffering in the world. Even if it is not totally avoidable could not the all-powerful, all-knowing God of naïve orthodoxy have made a better job of it? I strongly suspect, for reasons set out above, that the answer is a categorical *no*. The possibility of a world capable of supporting free individuals, tested and tempered by the uncertainties of life and destined for union in Christ seems to demand risk on the grand scale. Leibniz may have been right after all when saying that this was the best of all possible worlds. That is impossible to know, of course, but I have already given grounds for believing that complexity, which is essential to life, cannot exist without the potential for accident. There is certainly no ground for believing that any God worthy of the name could do better. We simply have no basis for such a conclusion, having no idea whether other worlds might be possible. I suspect the chief constraining factor responsible for this conclusion is the need to allow for free will. We value our free will above almost everything; our human dignity depends upon it and it is that which sets us apart from the rest of creation. But if *we* as individuals are free, then so is everyone else, and that means the risks created by their behaviour, foolish or otherwise, are unavoidable. To forego risk is to forego freedom; risk is the price we pay for our freedom.

The real question then is not why God chose to create this world but why he should have created anything at all. That is a big question for another day but in approaching it we might start by asking ourselves whether we would have preferred not to have existed. What we do know is that God did not exclude himself, in Jesus, from the human consequences of his choice to create this universe.

A CRITICAL ORTHODOXY

The path we have followed gives us the bones of a critical orthodoxy appropriate for an uncertain world. If my argument is correct, the threats to God's omnipotence, omniscience and providence are mistaken; the answer to the question 'Is God a risk taker?' is a qualified *yes*. This answer is not only in relation to his prime objective but in many of the secondary events and outcomes along the way. God's omnipotence thus remains intact because total control is simply not possible and God cannot do what is logically impossible. However our view of the matter is greatly enlarged when we glimpse the ingenuity in the interplay of chance and necessity. His way of working involves far greater subtlety than the crude mechanical analogy of naïve orthodoxy.

My response to the second question is that accidents are an inevitable consequence of there being a world sufficiently complex for life to exist. Hence suffering is unavoidable. God's omniscience is unchallenged because, although he knows all that can be known about the good and bad, the original act of creation carried with it certain implications which must have been recognised and could not be altered now without self-contradiction. God could and did take upon himself, in Jesus, the consequences of his decision to create in the first place.

His providence is to be seen in the rich potential with which the creation is endowed. The future is not wholly predetermined and hence is open to a measure of determination by God and ourselves. God's purposes are achieved as we align our actions with his will and, perhaps, also by his direct action.

All of this paints a picture on a canvas of breathtaking proportions beside which naïve orthodoxy, with which we started, appears unworthy of the God which nature and scripture reveals to us.

References

Ayala, Francisco, J. 2003. 'Intelligent Design: the original version', *Theology and Science* 1: 9–33

2007. *Darwin's Gift to Science and Religion*. Washington DC: The Joseph Henry Press

Barabási, Albert-László 2003. *Linked: How Everything Is Connected to Everything Else and What It Means for Business, Science, and Everyday Life*. New York: Plume, Penguin Group (USA) Inc.

Bartholomew, David J. 1982. *Stochastic Models for Social Processes* (third edition). Chichester, UK and New York: John Wiley and Sons Ltd

1984. *God of Chance*. London: SCM Press. Electronic version available at www.godofchance.com (no charge)

1996. *Uncertain Belief*. Oxford: Oxford University Press (paperback version, 2000)

1988. 'Probability, statistics and theology (with discussion)', *Journal of the Royal Statistical Society* A, 151: 137–78

Behe, Michael 1996. *Darwin's Black Box: the Biochemical Challenge to Evolution*. New York: The Free Press

Bernstein, Peter L. 1998. *Against the Gods: the Remarkable Story of Risk*. New York: John Wiley and Sons Inc.

Bovens, Luc and Hartmann, Stephen 2003. *Bayesian Epistemology*. Oxford: Oxford University Press

Box, Joan Fisher 1978. *R. A. Fisher, the Life of a Scientist*. New York: John Wiley

Brecha, Robert J. 2002. 'Schrödinger's cat and divine action: some comments on the use of quantum uncertainty to allow for God's action in the world', *Zygon* 37: 909–24

Buchanan, Mark 2002. *Nexus: Small Worlds and the Groundbreaking Theory of Networks*. New York and London: W. W. Norton & Company

Byl, John 2003. 'Indeterminacy, divine action and human freedom', *Science and Christian Belief* 15:2, 101–15

Coleman, Simon and Carlin, Leslie 2004. *The Cultures of Creationism: Anti-evolutionism in English-speaking countries*. Aldershot, UK: Ashgate Publishing Ltd

Colling, Richard G. 2004. *Random Designer: Created from Chaos to Connect with the Creator*. Borbonnais, Ill.: Browning Press

Conway Morris, Simon 2003. *Life's Solution. Inevitable Humans in a Lonely Universe*, Cambridge: Cambridge University Press

Davies, Paul 2006. *The Goldilocks Enigma: Is the Universe Just Right for Life?* Harmondsworth: Penguin Books

Dawkins, Richard 2006 [1996]. *Climbing Mount Improbable*. London: Penguin Books

Dembski, William, A. 1998. *Design Inference: Eliminating Chance through Small Probabilities*. Cambridge: Cambridge University Press

Dembski, William, A. 1999. *Intelligent Design: the Bridge between Science and Theology*. Downers Grove, Ill.: Intervarsity Press

2002. *No Free Lunch: Why Specified Complexity Cannot Be Purchased without Intelligence*. Lanham, Md.: Rowan and Littlefield

2004. *The Design Revolution: Answering the Toughest Questions about Intelligent Design*. Downers Grove, Ill. and Leicester, UK: Intervarsity Press

Dembski, William A. and Rüse, Michael (eds.) 2004. *Debating Design: From Darwin to DNA*. Cambridge: Cambridge University Press

Denton, Michael 1985. *Evolution: a Theory in Crisis*. London: Burnett Books, produced and published by Hutchinson

Diamond, Marion and Stone, Mervyn 1981. 'Nightingale on Quetelet', *Journal of the Royal Statistical Society* A, 244: Part I, 66–79, Part III, 332–51

Dowe, Phil 2005. *Galileo, Darwin and Hawking: the Interplay of Science, Reason and Religion*. Grand Rapids, Mich. and Cambridge, UK: Eerdmans Publishing Co.

Garon, Henry A. 2006. *The Cosmic Mystique*. Maryknoll, N.Y.: Orbis Books

Gleick, James 1987. *Chaos: Making a New Science*. Harmondsworth: Penguin Books

Gould, Stephen J. 1989. *Wonderful Life: the Burgess Shale and the Nature of History*. Harmondsworth: Penguin Books

Gregersen, Neils Henrik 1998. 'The idea of creation and the theory of autopoietic processes', *Zygon* 33: 333–67

(ed.) 2003a. From *Complexity to Life: On the Emergence of Life and Meaning*. Oxford: Oxford University Press

2003b. 'Risk and religion: towards a theology of risk', *Zygon* 38: 355–76

2006. 'Beyond secularist supersessionism: risk, religion and technology', *Ecotheology* 11: 137–58

Heyde, C. C. and Seneta, E. (eds.) 2001. *Statisticians of the Centuries*. Berlin: Springer

Hick, J. H. 1970. *Arguments for the Existence of God*. London and Basingstoke: Macmillan

Hodgson, Peter E. 2005. *Theology and Modern Physics*. Aldershot, UK: Ashgate Publishing Ltd

Holder, Rodney, D. 2004. *God, the Multi-verse and Everything: Modern Cosmology and the Argument from Design*. Aldershot, UK: Ashgate Publishing Ltd

Hoyle, Fred 1983. *The Intelligent Universe*. London: Michael Joseph

Hoyle, Fred and Wickhamasinghe, N. Chandra 1981. *Evolution from Space*. London: J. M. Dent and Sons

Kauffman, Stuart 1993. *The Origins of Order, Self-organization and Selection in Evolution*. Oxford: Oxford University Press

1995. *At Home in the Universe: the Search for Laws of Complexity*. London: Viking

2000. *Investigations*. Oxford: Oxford University Press

Kruskal, William 1988. 'Miracles and statistics: the casual assumption of independence', *Journal of the American Statistical Association*, 83: 929–40

Lipton, Peter 2004. *Inference to the Best Explanation* (second edition). London: International Library of Philosophy, Routledge

Lineweaver, Charles H. and Davis, Tamara M. 2002. 'Does the rapid appearance of life on Earth suggest that life is common in the universe?', *Astrobiology*, 2:2, 293–304

Matson, W. I. 1965. *The Existence of God*, Ithaca, N.Y.: Cornell University Press

Miller K. B. (ed.) 2003. *Perspectives on an Evolving Creation*. Grand Rapids, Mich. and Cambridge, UK: Eerdmans Publishing Company

Miller, K. R. 1999. *Finding Darwin's God: a Scientist's Search for Common Ground between God and Evolution*. New York: Cliff Street Books, an imprint of HarperCollins

Minsky, Marvin 1987. *The Society of Mind*. London: Heinemann

Monod, Jacques 1970. *Le hasard et la nécessité*. Paris: Editions du Seuil. English translation *Chance and Necessity* by Austryn Wainhouse (1972). London: Collins

Montefiore, Hugh 1985. *The Probability of God*. London: SCM Press

Morton, Glenn and Simons, Gordon 2003. 'Random worms: evidence of random and non-random processes in the chromosomal structure of the archaea, bacteria and eukaryotes', *Perspectives on Science and Christian Faith* 55: 175–84

O'Leary, Denyse 2004. *By Design or by Chance*. Minneapolis: Augsburg Books

Osler, Margaret J. 1994. *Divine Will and the Mechanical Philosophy: Gassendi and Descartes on Contingency and Necessity in the Created World*. Cambridge: Cambridge University Press (paperback, 2004)

Overman, Dean L. 1997. *A Case against Accident and Self-organisation*. Lanham, Md.: Rowman and Littlefield

Peacocke, Arthur 1993. *Theology for a Scientific Age: Being and Becoming – Natural, Divine and Human* (enlarged edition). London: SCM Press

Peterson, Gregory R. 2002. 'The Intelligent-Design movement, science or ideology', *Zygon* 37: 7–23

Primack, Joel and Abrams, Ellen 2006. *The View from the Centre of the Universe: Discovering Our Extraordinary Place in the Cosmos*. London: Fourth Estate

Polkinghorne, John 1984. *The Quantum World*. London: Longman
　　2006. 'Where is natural theology today?', *Science and Christian Belief* 18: 169–79

Pollard, W. G. 1959. *Chance and Providence*. London: Faber and Faber

Russell, Robert J., Murphy, Nancey and Peacocke, Arthur (eds.) 1995. *Chaos and Complexity, Scientific Perspectives on Divine*

Action, Vatican City State: Vatican Observatory Publications, and Berkeley, Calif.: The Center for Theology and the Natural Sciences.

Rüst, Peter 2005. 'Dimensions of the human being and of divine action', *Perspectives on Science and Christian Faith*, 57: 191–201

Saunders, Nicholas 2000. 'Does God cheat at dice?: divine action and quantum possibilities', *Zygon* 35: 517–544

　　2002. *Divine Action and Modern Science*. Cambridge: Cambridge University Press

Schrödinger, E. 1935. 'Die gegenwartige Situation in der Quanten-mechanik', *Naturwissenschaftern* 23: 807–12, 823–8, 844–9. Eng. trans. John D. Trimmer, *Proceedings of the American Philosophical Society* 124 (1980): 323–8

Sharp, K. and Walgate, J. 2002. 'The anthropic principle: Life in the universe', *Zygon* 37: 925–39

Smedes, Taede A. 2003. 'Is our universe deterministic? Some philo-sophical and theological reflections on an elusive topic', *Zygon* 38: 955–79.

Smith, Peter 1998. *Explaining Chaos*. Cambridge: Cambridge University Press

Sproul, R. C. 1994. *Not a Chance: the Myth of Chance in Modern Science and Cosmology*, Grand Rapids, Mich.: Baker Books

Strogatz, Steven 2003. *SYNC; the Emerging Science of Spontaneous Order*. London: Penguin Books

Swinburne, Richard 2004. *The Existence of God* (second edition). Oxford: Clarendon Press

Van Till, Howard 2003. 'Are bacterial flagella designed: reflections on the rhetoric of the modern ID movement', *Science and Christian Belief* 15: 117–40.

Ward, Keith 1990. *Divine Action*. London: Collins

　　1996. *God, Chance and Necessity*. Oxford: OneWorld Publications

　　2000. 'Divine action in the world of physics: response to Nicholas Saunders', *Zygon* 35: 901–6

Wildman, Wesley J. 2004. 'The Divine Action Project', *Theology and Science* 2: 31–75

　　2005. 'Further reflections on "The Divine Action Project"', *Theology and Science* 3: 71–83

Woolley, Thomas 2006. 'Chance in the theology of Leonard Hodgson', *Perspectives on Science and Christian Faith* 56: 284–93

Further reading

Most of the existing books relating to chance and purpose have been referred to in the text and are listed in the References. Some, however, have been omitted because they have been superseded by more recent books covering the same point and others because they deal with the same issues in a more general way or from a different perspective. An exhaustive listing of those in the latter category would be prohibitively long but the following short list includes some of the most important. In the main, these books come from the science side and a few provide technical background at a popular level.

Brooke, John Hedley 1991. *Science and Religion: Some Historical Perspectives*. Cambridge: Cambridge University Press

Conway Morris, Simon 1998. *The Crucible of Creation: the Burgess Shale and the Rise of Animals*. Oxford: Oxford University Press

Davies, Paul C. W. 1982. *The Accidental Universe*. Cambridge: Cambridge University Press

Eigen, Manfred and Winkler, Ruthild 1982. *The Laws of the Game: How the Principles of Nature Govern Chance*. London: Allen Lane, Penguin Books Ltd

Haught, John F. 2003. *Deeper than Darwin: the Prospect for Religion in the Age of Evolution*. Boulder, Colo.: Westview Press

Peacocke, Arthur, 2004. *Creation and the World of Science: the Reshaping of Belief*. Oxford: Oxford University Press (first published in 1979)

Peterson, Ivars, 1998. *The Jungles of Randomness: a Mathematical Safari*. London: Penguin Books

Prigogine, Ilya and Stengers, Isabelle 1984. *Order out of Chaos: Man's Dialogue with Nature*. London: Heinemann

Ruse, Michael 2004. *Can a Darwinian be a Christian?: the Relationship between Science and Religion Today*. Cambridge: Cambridge University Press

Southgate, Christopher (ed.) 2005. *God, Humanity and the Cosmos: A Companion to the Science–Religion Debate* (second edition, revised and enlarged). London and New York: T. & T. Clark International

Stannard, Russell (ed.) 2000. *God for the 21st Century*. Philadelphia: Templeton Foundation Press, and London: SPCK

Stewart, Ian 1989. *Does God Play Dice?: the Mathematics of Chaos* Oxford: Basil Blackwell

Watts, Frazer (ed.) 2007. *Creation: Law and Probability*. Aldershot, UK: Ashgate Publishing Ltd

Index

observations, paradoxical effect of
making, 142, 143–50
probability theory and, 143, 151
Schrödinger's cat, 143–50
statistical approach to, 136–37,
143–50
superposition of multiple events, 142,
143–50
uncertainty in, 137, 151
wave function in, 139, 143, 147
Quetelet, Adolphe, 119n2, 194n11
queueing theory, 161n2, 161–2

radioactive emissions
God's action in random events,
216–17
God's choice of chance, 218
integral part of created order,
challenges to chance as, 198
normal law, 39
Poisson's law, 36, 218
statistical laws, 131–2
total unpredictability, impossibility
of, 221–2
railway built on bog, 233
random chemistry, 163–4
random nets, 46–9
random numbers, genuine, 59
random sampling, 38, 164–6
Rappoport, Anatol, 46
rationality, chance not opposed to, 172
rejection sets and significance testing
Intelligent Design and, 100, 102,
103–6, 107–9
very small probabilities, 85–7
relative probabilities and inference to
the best explanation, 91–5
Rényi, Alfréd, 44
Rhinehart, Luke, 128n3
risk, 5, 223–42
benefits of, 229–31
critical orthodoxy, development of,
231, 241–2
free will and, 239–40
God as risk-taker, 223–7
Incarnation of Jesus and, 228, 237
nature of belief and, 231–5

sovereignty of God and, 226, 235
theology of, 235–41
world, theological implications of
riskiness of, 227–9
rumours and gossip, 48–9
Ruse, Michael, 100
Russell, Robert John, 151
Rüst, Peter, 155n8

sampling, random, 38, 164–6
SARS, 49
Saunders, Nicholas, 137, 138, 140
scale-free networks, 50
scales, *see* levels and scales
Schrödinger's cat, 143–50
science
biblical view reinforced by lawfulness
of, 226
Intelligent Design as, 98, 114–15
theology, scientific inference as
model for, 232–5
SDA (special divine action), 138
sex ratios, 31–2, 79, 87, 103–5
SIDS (cot deaths) and probability
theory, 68–9, 71
significance testing
Intelligent Design and, 100, 102,
103–6, 107–9
very small probabilities, 85–7
'six degrees of Kevin Bacon' game,
43
Sliding Doors (film), 128
Small Numbers, Law of, 34
'small world' phenomenon, 43–5
snooker, 230
social order, role of chance in, 192–5
sovereignty of God, 5
accidents and coincidences, 58
chance and, 99
integral part of created order,
challenges to view of chance as,
196–7, 200, 203–4
risk and, 226, 235
statistical laws and, 125
special divine action (SDA), 138
specified complexity and Intelligent
Design, 106, 108–9, 112